Generative AI Agents: Foundations and Applications

Rajiv Kumar

Made with ♥ on the Notion Press Platform

www.notionpress.com

Dedication

To my family, whose unwavering support has been my foundation in both personal and professional life.

To the pioneers of Artificial Intelligence, whose work has paved the way for the breakthroughs we continue to see today.

To my colleagues and mentors at Oracle and beyond, whose collaboration and innovation fuel the drive for excellence in AI.

This book is dedicated to everyone committed to shaping a future where AI transforms industries and enriches lives

Contents

Preface.. 9

Acknowledgment 11

Part I: Foundations of Generative AI Agents

Chapter 1:   Introduction to Generative AI Agents................**15**

1.1   What are Generative AI Agents? 18

    1.1.1   Definition of Generative AI....................................19

    1.1.2   Characteristics of Generative AI Agents...............22

1.2   Evolution from Static Models to Autonomous Agents......23

    1.2.1   Overview of Traditional AI Models
        and Generative AI Agents 25

    1.2.2   When to Combine Generative AI
        with Traditional AI 29

    1.2.3   Modern Generative AI Agents 31

1.3   Real-World Applications of Generative AI 32

    1.3.1   Real-World Use Cases for Generative
        AI Agents 32

    1.3.2   Real-World Application for Generative
        AI Agents 38

Multiple Choice Question (MCQs) 41

Chapter 2:   Core Components of Generative AI Agents **44**

2.1   Generative Models as the Brain 44

2.2   Knowledge and Memory Systems 51

2.3   Decision-Making Capabilities...55

2.4   Integrating Neural Networks in AI Systems 59

Multiple Choice Question (MCQs) 67

Chapter 3: Ethics and Safety in Generative AI Agents..........**70**
3.1   Preventing Misuse of AI 70
3.2   Bias and Fairness in AI Systems......................................75
3.3   Ethical AI: Developing Safe and Responsible Agents........82
3.4   Transparency and Accountability in AI Systems 87
Multiple Choice Question (MCQs) 90

Chapter 4: How Generative AI Agents Work.....................**93**
4.1   Perception: Gathering and Processing Input 96
4.2   Generation: Producing Outputs and Actions.................. 101
4.3   Interaction: Engaging with Users and Systems 103
4.4   Challenges in Improving AI Efficiency 107
Multiple Choice Question (MCQs) 111

Part II: Building Generative AI Agents

Chapter 5: Frameworks and Architectures........................**117**
5.1   Agent Architectures: Core Concepts 117
5.2   Retrieval-Augmented Generation (RAG)......................... 124
5.3   Modular Approaches to AI System Design 129
5.4   AI Frameworks for Real-World Applications 131
Multiple Choice Question (MCQs) 135

Chapter 6: Memory and Knowledge in Agents **138**
6.1   AI Agent Memory 138
6.2   Short-Term vs. Long-Term Memory in AI 140
6.3   Using Knowledge Graphs for Knowledge
       Representation.. 147
6.4   Strategies for Managing Knowledge
       in Evolving AI Systems 152
Multiple Choice Question (MCQs) 157

Chapter 7: Training and Fine-Tuning Generative AI Agents 160

7.1   Pre-Training and Fine-Tuning Basics 160
7.2   Reinforcement Learning for Advanced Capabilities 167
7.3   Transfer Learning for Industry-Specific Use Cases 171
7.4   Fine-Tuning Techniques and Optimization
      Challenges 176
Multiple Choice Question (MCQs) 180

Chapter 8: Enhancing Agent Capabilities 183

8.1   Natural Language Understanding and Generation 183
8.2   Multi-Modal Capabilities in AI Agents 189
8.3   Training for Industry-Specific Skills 193
8.4   The Interplay of Language, Context, and
      Task Performance 199
Multiple Choice Question (MCQs) 201

Part III: Applications of Generative AI Agents

Chapter 9: Conversational Agents and Chatbots 207

9.1   The Evolution of Chatbots: From Scripts to AI 209
9.2   Building Chatbots for Specific Industries 215
9.3   AI Agents for Customer Support and Service 219
9.4   Case Studies: Success Stories in Chatbots 223
Multiple Choice Question (MCQs) 227

Chapter 10: Creative Agents 230

10.1   AI am in Content Creation 230
10.2   Storytelling and Gaming Applications 234
10.3   Creative AI in Art, Music, and Literature 238
10.4   The Role of AI in Collaborative Design 242
Multiple Choice Question (MCQs) 247

Chapter 11: Productivity and Automation Agents 250

11.1 Coding Assistants for Developers 250

11.2 AI in Business Process Automation 254

11.3 Tools for Team Collaboration and Workplace
Productivity 259

11.4 Personal Productivity Agents.................................... 263

Multiple Choice Question (MCQs) 268

Chapter 12: Autonomous Decision-Making Agents 271

12.1 AI Agents for Reasoning and Planning........................ 271

12.2 Optimizing Decision-Making in Complex Scenarios..... 275

12.3 Applications in Strategic Business Decisions 283

12.4 The Role of AI in Healthcare Decision Making 286

Multiple Choice Question (MCQs) 290

Part IV: Advanced Topics in Generative AI Agents

Chapter 13: Scaling Generative AI Agents 295

13.1 Infrastructure Considerations for Scaling AI 295

13.2 Distributed Systems for Large-Scale Agents 299

13.3 Managing Latency and Real-Time Performance 303

13.4 Cost and Resource Optimization 308

Multiple Choice Question (MCQs) 316

Chapter 14: The Future of Generative AI Agents.................319

14.1 Towards Artificial General Intelligence (AGI)................ 319

14.2 Case Study: AI in Content Creation............................. 324

14.3 Transforming Industries with AI Agents 327

14.4 Ethical and Social Challenges of AGI 330

Multiple Choice Question (MCQs) 335

Bibliography...339

About the Author..351

Preface

Artificial intelligence is fast-moving to become the next big thing, and generative AI agents are being heralded as the game-changer across all industries. Powered by state-of-the-art neural networks, decision-making systems, and generative models, these agents are a new form of intelligence: one that can perceive, reason, and act autonomously. This book, Generative AI Agents: In Foundations and Applications, we attempt to examine the foundations and the architecture, as well as the real-world applications of this groundbreaking technology.

This book is inspired by the increasing need to know how these agents work, what challenges they present, and what opportunities they offer. Generative AI agents are changing the way work gets done from customer support to creative design, driving efficiencies and creating innovation across domains. But as their potential increases, so do the ethical, social, and technical questions about their development and deployment.

This book is structured into four parts: The process of building AI agents, the foundational concepts, applications in various domains, and the advanced topics for scaling and future development. This book is designed for AI practitioners, researchers, business leaders, and everyone curious about the mechanisms and implications of generative AI agents. This book will help you whether you are building these systems, tapping their power, or identifying their effects, to navigate this fast-moving landscape.

I hope this work will be a good starting point for you to become inspired by the power of generative AI to transform and responsibly use it.

Acknowledgment

Without the support, guidance and contributions of the many individuals and organizations, this book would not have been possible. I wish to express my deepest gratitude to the artificial intelligence mentors and colleagues who have pushed me to take this work seriously and think deeply about the possibilities and the responsibilities of AI technology.

I'd like to thank the research community, whose groundbreaking work on generative models, neural networks, and AI architectures formed the basis of the concepts described in this book. I want to thank my peers and collaborators for their discussions, feedback and encouragement as I worked through this process.

I want to thank my family and friends for their constant support and encouragement throughout the very long hours I dedicated for researching and writing.

I would also like to thank the broader AI and technology ecosystem, which continues to be a driver of innovation and the bounds of what is possible. My hope is that this book adds constructively to the ongoing conversation about what generative AI agents can mean for the future.

PART I

Foundations of Generative AI Agents

CHAPTER 1

Introduction to Generative AI Agents

The world as we know it is changing rather fast and there is one concept which for sure has the potential to revolutionize our future and that is generative AI. Feel like getting out of the bed to a world where an intelligent system, with your permission, has selected better news than what you want to read in the morning or a system where AI can delve through millions of records to recommend the best diet for you (Mariani and Dwivedi 2024).

Generative AI (GAI) is a technology with high research interest as well as practical applications across many domains. This is done possible by using many resources assumed during its pre-training phase and is in its pre-training proposed. It is insightful to use an example of how large language models (LLMs) are progressively being built, specifically the Generative Pre-trained Transformer from OpenAI. However, as all-around models, LLMs can sometimes be ineffective while providing answers to highly specific questions, limited by the range of their pre-training. Not surprisingly, this becomes a challenge and to solve it, there is a method known as Retrieval Augmented Generation (RAG) that helps to enhance the model to provide particular responses from the use of external databases. Thus, cohering with the formulation of the generative AI agent, LLM and RAG technologies synergistically complement each other. The generative AI agent is a very efficient and smart tool that is used to generate adaptive and personalized content in given fields using unique outside sets of data.

Retrieval Augmented Generation (RAG) is a paradigm that consolidates the benefits of large language models (LLMs) and external knowledge bases to create a new transformative framework. However, LLMs like GPT or BERT excel at understanding and producing natural language, but they tend to produce outputs within the pre-trained data, which limits their reality, making them vulnerable to errors in dealing with specific or real-time information. To alleviate these problems, RAG augment's generative ability with a retrieval component that allows integrating external, domain-specific knowledge into the response generation process.

First, the user query is received by RAG framework, which then processes the query and uses a retrieval module to find relevant documents or information in an external database. For this reason, it can take advantage of advanced tools such as Elasticsearch, FAISS, etc. to do a very precise retrieval of highly relevant data. After retrieved, the documents are sent to the LLM to be enriched with the input query and passed as context. The augmented context enables the model to output responses grounded in both context and fact, addressing the limitations of traditional LLM outputs.

RAG is one of the key advantages of being able to dynamically adapt to domain-specific and time-sensitive requirements. For example, in cases like customer service, RAG systems can fetch policy documents or FAQs relevant to a company and provide an accurate and personalized response. In the same vein, the framework can be used in research or legal application to integrate the most recent academic papers as well as case laws to produce outputs that are highly specialized and informed. This adaptability applies to industries such as healthcare, finance, and education, where having the right and up-to-date information is necessary.

RAG also excels at making factual correctness of responses based on external data. RAG systems differ from static models that only use pretrained models in that they can access real-time information to maintain the relevance and trustworthiness of the responses. RAG's dynamic capability, however, is very valuable in situations where updates must remain current, like financial market analysis, breaking news summaries, or emerging trends in technology

RAG has been used in practice in many applications. A RAG system versus, if, for example, a user would want to know recent trends in generative AI. In the retrieval module, current research papers or news articles would be retrieved by LLM and passed on to it. The final response would be the output of a synthesis of the model's generative capabilities and the insights from the retrieved documents. By integrating retrieval and generation, the depth of responses improves as well as building trust with users through verifiable and rounded information. RAG, which emerges as a robust and versatile framework by combining the creative power of generative models with the precision of the retrieval systems. Key to its ability to deliver timely, domain-specific, and accurate responses makes it key innovation in the growth of AI, and can be used for many such applications in a range of industries. This opens up the possibility of AI beyond static pre-training and evolving into more dynamic and responsive knowledge synthesis and decision-making tool.

RAG has been used in practice in many applications. A RAG system versus, if, for example, a user would want to know recent trends in generative AI. In the retrieval module, current research papers or news articles would be retrieved by LLM and passed on to it. The final response would be the output of a synthesis of the model's generative capabilities and the insights from the retrieved documents. By integrating retrieval and generation, the depth of responses improves as well as building trust with users through verifiable and rounded information.

RAG, which emerges as a robust and versatile framework by combining the creative power of generative models with the precision of the retrieval systems. Key to its ability to deliver timely, domain-specific, and accurate responses makes it a key innovation in the growth of the AI, and can be used for many such applications in a range of industries. This opens up the possibility of AI beyond static pre-training and evolving into more dynamic and responsive knowledge synthesis and decision-making tools (Z. Wang et al. 2024).

1.1   What are Generative AI Agents?

A recent advancement in artificial intelligence is the creation of generative AI agents, which use complex formulae derived from machine learning concepts to create novel and distinctive functional content that can create material, make decisions, and communicate with people. These generative AI agents function as intricate mathematical models that generate new ideas, judgements, and even works of art from large amounts of data. Generative AI agents are being incorporated into more and more businesses as technology advances, improving innovation and productivity.

A generative AI agent is an enhanced artificial intelligence within the generation instance of its capability, integrated with other model functionalities such as data analysis, decision-making, and action. These agents, unlike many other generative model AI, are capable of planning and executing tasks as well as reviewing the output as needed and incorporate new data, and for these reasons they are incredibly useful in many respects. But this is where generative AI agents stand out since such AI can create content and also perform actions on it independently. They can gather information, decide what it means and take particular actions relating to it. Such an approach helps them to finalize multifaceted work, starting with data preliminary processing and ending by the final task performance; thus, the ability to work independently and to solve problems in the conditions of uncertainties is shown.

Generative AI agents have the potential to revolutionize sectors like healthcare, banking, and customer service by fusing generative capabilities with autonomous operations. This will allow them to improve overall efficacy, accuracy, and efficiency. Understanding these agents is key to leveraging their full potential in developing innovative solutions across various domains (Pattam 2024).

1.1.1   Definition of Generative AI

Generative AI is a kind of machine learning (ML) model that can produce new text, graphics, music, and code by using the knowledge it has gained from the examples it has been given. These models pick up knowledge through training, which is the process of seeing and matching patterns. Information databases and deterministic information retrieval methods are not what generative AI models are. This is because they are prediction engines and as such, depending on which outputs they are being programmed to look for, they can come up with different results yet they are responding to the same input.

A generative AI agent is an AI system intended to create new knowledge and data from obtained patterns of information. Instead of specifying deterministic structures as in the case of traditional AI applications, generative agents produce new information, for example:

Text, music, pictures, and other types of data might all fall under this category. Generative AI is frequently used in the following applications: Figure 1.1.

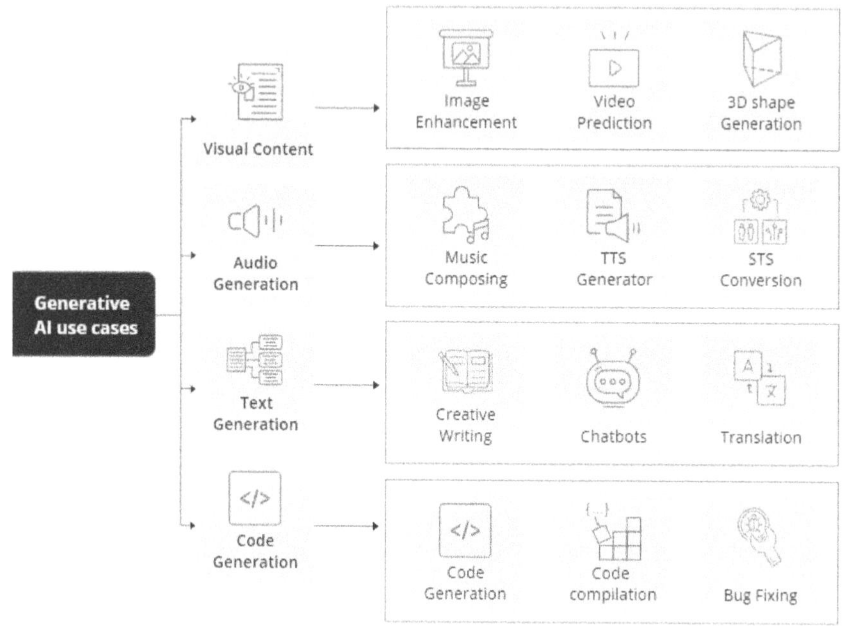

Figure 1.1: Use cases of Generative AI.

Source: − (Akash Takyar 2022)

- **Image generation:** Generative AI means that products of art can be created based on producers; reproduced and new images can be created such as aesthetics, new art pictures and realistic pictures for usage in advertisements, and movies among others.
- **Music generation:** Words such as generative AI and generative model in music can also be implemented to develop new music because they analyze different patterns and trends in music and produce a new tune or pattern.
- **Text generation:** It is identified that with the help of generative approaches, text patterns and trends can be analyzed to create new items in the text such as new sentences or paragraphs.

- **Code generation:** Through the provision of a simple language description of what is expected of the code after the later preparation, the model may be used to compile code. As an example, you might provide the model a natural language description of what you actually want it to produce such as creating a button with text 'Click Me' if it was to produce code for a new button on a web page. The model would then generate the code which should be used to construct the button.

In order to find innate systems and patterns in large data sets, these generative AI agents use deep learning as a machine learning technique. They are able to provide outcomes that are situation-appropriate and relevant in this way. Models such as transformer-based topologies or Generative Adversarial Networks (GANs) are frequently used to fuel generative AI agents, enabling them to learn from enormous volumes of data and provide superior outcomes.

Generative AI models when integrated with plugins or APIs extend their range of applicability and importance namely through these agents. For instance, the ability to link a chatbot to calendar APIs will make it a virtual assistant with the ability to schedule a meeting on its own.

In order to generate fresh and unique content, generative AI architecture use neural network models to find patterns and structures in existing data. Using various learning techniques, such as unsupervised or semi-supervised learning, for training is one of the innovations of generative AI models. As a result, companies may now more simply and quickly use vast volumes of anonymous data to build baseline models. Base models can serve as the foundation for AI systems that can accomplish a variety of functions, as the name implies.

1.1.2 Characteristics of Generative AI Agents

Generative AI agents are systems capable of autonomously performing tasks by leveraging generative AI models.

- **Creativity:** They can create new materials, text, images or audio based on patterns learnt from the training set.
- **Adaptability:** They build up from experiences and refine outputs to be in harmony with existing consumer trends and certain situations.
- **Autonomy:** Like other self-motivated AI advisors, they are self-acting but inside the code parameters given to them, though extra creative regarding selection.
- **Context Awareness:** They are able to comprehend, extrapolate to different situations, and provide more appropriate and sophisticated responses.

Challenges and Risks

Although the ability to use Generative AI agents to improve performance and increase revenue is rather great, there are still a number of difficulties and threats to overcome:

- **Data Privacy and Security:** Being that AI agents work with special data, data privacy and security come first. Organizations are, therefore, required to incorporate effective insurance measures to mitigate the probability of leakage of data or unauthorized access.
- **Bias and Fairness:** AI contains bias at the time it was trained and, as such, can replicate the same bias in its judgment. Addressing bias and the problems of biased results affecting individuals using AI systems is important to aggressiveness and afford individual customers equal treatment.

- **Explainability and Transparency:** AI agents work in an opaque manner. Hence, one needs to know what was the decision-making process that was followed. To increase the trust and take responsibility for using AI, there must be increased focus on the explain ability and transparency of AI.
- **Regulatory and Ethical Considerations:** The use of AI agents presents several questions of ethics and regulation. Such issues are an increasing concern due, to the increasing deployment of AI applications, and for this reason, creating ornate frameworks on how AI should be regulated is critical.

Differentiation from Traditional AI

Conventional systems employing artificial intelligence for planned and designed tasks to an extent meet these requirements but generative AI agents are more versatile and adaptive. Static system is rigid by nature for they only form static frame work which they have to follow and on the other hand generative agents are effective in dynamic world. These are important characteristics for the system to facilitate applications – which require context sensitivity and information. The transition from a physical model to an independent self-acting entity is an exciting transition in artificial intelligence (AI) innovation owing to evolution in machine learning, generative models and integration.

1.2   Evolution from Static Models to Autonomous Agents

The shift from 'static models' to 'autonomous agents' in AI can be thought of as a leap where systems progress from executing specific, pre-specified tasks based on input data to actively sensing, deciding and responding to their environment in real time and principally without reference to a human operator to accomplish objectives; or from the state of data processing to that of intelligence.

1. **Static Models**: There were two kinds of AI systems initial, which ranged from static models that included basic rule application or strictly programmed algorithms to perform certain operations. These models were somewhat restricted, for they could only accept input information and produce an output in a very narrow framework. They were not flexible, logical or able to think through what to do when the fighting started. For instance, the first generation of spam filters sorted emails into categories using code imprinted on it, meaning that the filter could not learn on its own the new methods being used to spam the mass without human interference.

2. **Generative Models:** The use of generative AI models will be considered a breakthrough. Shared intelligence concepts, such as GPT (Generative Pre-trained Transformers) and DALL·E and provided an option to generate dynamic results, which may include natural language text, images, or music given the scope of the trained datasets. These models could interpret context and this produced creative and human like responses. Nevertheless, while they were great at creating content, they were just tools and programs that needed to be manually led on to do tasks and could not do them on their own.

   ➢ **Autonomous Agents:** The New Era of Modern generative AI agents extend the generative model archetypes by incorporating the element of reasoner, memory and decision-maker. Many of these agents are no longer simply content-generating—they are capable of planning, carrying out, and self-adjusting tasks from end to end. Key enablers of this transition include:

- **Reasoning Capabilities:** Allow agents to evaluate complex scenarios and deduce optimal solutions.
- **Memory Systems:** Enable them to retain context over time, improving performance in multi-step or ongoing tasks.
- **Decision-Making Abilities:** Empower agents to analyse multiple options and act based on the most effective approach.

➢ **Integration with Tools and APIs**

Autonomous agents utilize APIs, plugins, and external tools to enhance their functionality. For example, an AI agent might integrate with a calendar API to schedule meetings or use financial tools to analyze market trends and make investment recommendations. This interconnectedness allows them to perform tasks beyond content generation, such as data analysis, process automation, and real-time decision-making.

1.2.1   Overview of Traditional AI Models and Generative AI Agents

In this section to overview of Tradition AI model and Generative AI agents' model to discuss limitations, advantages and Key difference of both model in below:

Traditional AI: A Brief Overview

In order to accomplish a job, traditional AI, also known as narrow or weak AI, needs intelligence. It talks about systems that react to certain inputs. Based on data, these systems are able to forecast or make decisions. Consider engaging in computer chess.

Knowing every regulation, the computer can anticipate and move according to a plan. It chooses from pre-programmed chess strategy. Traditional AI is like a great strategist who can make clever decisions within rules. Siri and Alexa, Netflix and Amazon recommendation systems, and Google's search algorithm are other typical AIs. These AIs are trained to follow rules, do a task well, but they do not develop anything new.

Traditional AI is sometimes referred to as an expert system or rule-based AI. AI is a kind of artificial intelligence that uses preset rules and algorithms to carry out certain tasks and make decisions. The goal of conventional AI is to be exceptionally skilled at one or a limited number of activities, as opposed to general AI, which strives to demonstrate human-like intelligence across a range of tasks.

> **Advantages of Traditional AI**

- **Transparency:** the decision-making procedures are far more straightforward and easier to understand
- **Reliability:** Traditional AI produces consistent and predictable outcomes with the same input data.
- **Domain-Specific Expertise:** Traditional AI works well in trades with deep knowledge and accurate decision-making.

> **Limitations of Traditional AI:**

- **Limited Adaptability:** Traditional artificial intelligence may fail to work dynamically in this regard because a change in rules must be made manually.
- **Scalability:** Scalability is always an issue for traditional AI systems because rules need to be manually acquired and managing a large number of rules is impracticable.

- **Lack of Generalization:** These systems are unable to put the knowledge into use other than the rules given, and hence not as flexible.

Generative AI: The Next Frontier

In its widest sense, generative AI, also known as generating AI, is sometimes referred to as the second generation of AI. Strong AI, sometimes referred to as generative AI, is the kind of AI that can produce text, video, graphics, and other types of material directly. Thankfully, generative AI has become more and more popular. It learns from past training data and is now widely used. It can open and produce not just text but also images, music, and even code. When given a collection of data, generative AI models learn the training set's probabilistic distribution to generate new data.

- ➢ **Advantages of Generative AI**

  - **Creativity:** Generative AI is used to learn data patterns so as to generate new data. It can be used to generate pictures, translate languages and obtain suggestions on information.
  - **Adaptability:** In Generative AI, data and settings changes do not require the reinforcement rules in order to adapt. More data can make it better.
  - **Generalization:** One capability of generative AI is that it can transfer its knowing to be able to perform several tasks in different domains that are related to each other.

- ➢ **Limitations of Generative AI**

  - **Complexity:** Generating and especially fine-tuning Generative AI models requires much data and powerful systems.

- **Lack of Transparency:** Most deep learning models cannot be easily explained this makes it difficult to determine accountability or transparency.
- **Ethical Concerns:** The tools of Generational AI bring ethical issues in front of the world such as deep fake content and misuse.

The Key Difference:

It is also possible to describe the difference between the regular and generative AI as the one between their effectiveness and the sphere of their usage. While the traditional AI systems are standards based on the data to make prediction, the generative AI has an added feature of generating a fresh array of data similar to the training data. Key difference: Presents Table 1.1.

Table 1.1: Traditional AI vs Generative AI: Key Differences.

Aspect	Traditional AI	Generative AI
Primary objective	Performs predefined tasks based on rules	Produce new, original content or data
Strengths	Capable of effectively interpreting and performing specific responsibilities	Innovative, creative, and adept at managing ambiguity
Weaknesses	May lack creativity and innovation, less adept at handling uncertainty	Not as good as conventional AI systems for identifying patterns and solving task-specific problems

Key features	Decision trees, pattern recognition, predictive modeling	Deep learning, neural networks. creative data generation
Data requirements	Depends on labeled/structured data for training	Uses large structured and unstructured datasets for training
Technological approach	Structured analysis and logical processes	Dynamic, creative, and capable of generating innovative outputs
Transparency	Structured analysis and logical processes	Can be less transparent due to complex learning algorithms, making it challenging to understand how particular outputs are derived
Applications	Predictive analytics, fraud detection, personalized recommendations, process automation, decision support systems	Automated content generation, AI-generated art, synthetic data generation, content moderation.

Source: – (Nguyen 2024a)

1.2.2   When to Combine Generative AI with Traditional AI

The distinction between generative and classical AI should be made clear. They can work in tandem to achieve the intended business goal in certain business settings. For example, we obtain the final result when the generative AI model receives output from the conventional model as part of a text prompt. The following are some use cases that make sense when utilizing both traditional and generative AI capabilities:

- Conventional AI can forecast client turnover by examining historical data. Integrating this information with an LLM or generative AI-powered chatbot lets your sales staff test forecasts via natural language chats. Business intelligence (BI) dashboards can be created via chatbot discussion.

- Conventional predictive AI can anticipate the risks associated with a particular use case, whereas generative AI can model several situations to assist in developing potential mitigation techniques.

- Conventional predictive AI is capable of identifying consumer categories to assist in the development of tailored campaigns and marketing. Then, using generative AI, you can create customized marketing content for every section that has been discovered.

- In order to convert video data into text, traditional AI computer vision can recognize and categorize sign language. Before producing an accurate translation in written form, including translation in several languages, current and future sophisticated generative AI can assist in taking into account contextual and other minor distinctions within sign language. Frequently, generative AI may provide output in speech format after translating text to sign language, allowing signing and non-signing people to have a complete discussion.

- In order to extract useful aspects from videos, conventional AI may analyze videos and include video analytical features. It can, for example, provide text extraction, object detection, human detection, and video resource discovery. With such findings, generative AI may then produce new experiences, such as chatbots, listings, reports, or articles.

1.2.3 **Modern Generative AI Agents**

A "modern generative AI agent" refers to an advanced artificial intelligence system that combines the capabilities of generative AI models (like GPT) with autonomous functionalities, allowing it to not only generate new content but also actively make decisions, interact with users, and execute actions based on its analysis of data, essentially acting as an independent agent with the power to create new information and complete complex tasks across various domains.

Current generative AI agents are sophisticated self-contained systems that incorporate other human mental faculties like inference, storage, and choice. They make an easier task of solving intricate problems and performing under different possibilities than the typical AI equipment. Let's break this down:

- Thus, there are three fundamental aspects in the present generative AI agents, which makes them exceptionally effective. First, thinking includes the ability to understand a problem, to assess information, and deduce, rational consequences. This enables them to task complex operations which simple solutions cannot accommodate all of them. For instance, the health care domain of an AI agent is capable of analyzing symptoms and compared these with data in medical database to provide possible diagnosis or treatment plans.
- One is evidence that memory allows for storage and retrieval of information of related agents or from such past interactions or incidences. Of these, this skill enables one to retain context and therefore enable an efficient, long-term, task-oriented performance tuned to the individual. A customer care chatbot can recall a user's previous inquiries and provide personalized results which will increase satisfaction.

- Third, decision-making helps to make agents evaluate several possibilities and select the most efficient action. This is necessary in areas where priorities clash or its outcome cannot be ascertained. An example instance is a finance artificial intelligence digital agent, it processes real-time information and gives the user the best way to invest. These integrated attributes enable generative AI agents to self-perform tasks and optimally work in changing conditions.

1.3 Real-World Applications of Generative AI

Global industries are adopting generative artificial intelligence (AI) and its growing organizational solutions and creative development approaches. Some examples of use case models are; Generative AI Architecture for generating complex systems and Generative AI Solutions – private LLMs for building individual language learning models in secure settings. In business, Generative AI in contact Centre refines pattern of customers' behaviours, Generative AI Search helps in enhanced information seeking.

Healthcare and banking use AI for data analysis and personalised recommendations. Generative AI provides effective digital marketing campaigns, while AI promotes urban planning and efficiency in Smart Cities. Generative AI improves retail inventory planning and customer service. With Pilot Generative AI exhibiting early developments, Generative AI platforms as well as technologies support these applications.

1.3.1 Real-World Use Cases for Generative AI Agents

Generative AI agents are transforming industries by leveraging advanced models to create content, simulate scenarios, and enhance decision-making are some noteworthy examples of real-world applications. Examples of real-world use cases for generative AI agents, such as coding, creation, and security.

1. Creation

The following are a some of the most widely used generative AI applications for content production:

- **Copy and content creation:** Programs like as Writer and Jasper AI provide copy templates to help create new message concepts, scripts, closed captioning, product descriptions, and frequently asked questions.
- **3D models and art:** Programs such as DALLE-2 and Midjourney may generate images and artwork to help speed up creative ideas. Enterprises employ ChatGPT integration services such as DALLE, Midjourney, and Stable Diffusion to create visual content for social media these days. When converting words into pictures, DALLE, for example, may assess up to 12 billion elements and generate realistic graphics and artwork from text descriptions.
- **Translation:** Generative AI models are capable of translating text between languages quickly and precisely.
- **Marketing and advertising:** It is possible to create marketing and advertising materials such as catchphrases, ad text, product descriptions, and content for social media promotion.

2. Coding

Software developers would be much more productive if they used generative AI since it would enable them to swiftly transition between multiple programming languages, understand a variety of tools and approaches, automate code development, anticipate and prevent problems, and maintain system documentation.

- **Code generation:** GPT Applications programs can produce code from natural language descriptions by training a model on a huge dataset of text and code. After preparation, the model may construct code if it is given a clear description of what the code should do. A natural language description such as "create a button with the text "Click Me" to generate code that adds a new button to a web page" might be sent to the model. The model then produces button code.

- **Debugging code:** One way of applying generative AI in order to debug code is to train a model with a large collection of code and known defects first. Subsequently, the model may again be used to course check the code in question by inputting this code into it. Next, the model will generate the likelihood of errors that could be the cause of the code's fault. With that, you might use this list to search the bugs on your code and fix them.

- **Testing code:** It may be tested by first training on a sizable code dataset and then utilizing generative AI experiments. When the model is ready, you may utilize it to do experiments with your code in the following ways. reference codes for a free Excel spreadsheet resume. The model will generate unique test cases that evaluate every feature of the code. After that, you may use them to test your code and find errors.

- **Documenting code:** The generational AI can program a computer after going through a big set of code and comment. When ready, the model can be used to write comments and other documentation on the code's usefulness. This documentation can help you understand the code and make updating and improving it easier.

- **Teaching coding:** Generative AI may teach coding by using a large code and instructional dataset to train a model.

When the model is ready, it produces simple exercises and activities that aid students in learning new coding ideas. To educate students how to develop a button for a website, you might utilize the model to make a lesson. Interactive tutorial tasks let students apply their knowledge.

3. Security

Generative AI will eventually support enterprise governance and information security, prevent fraud, improve regulatory compliance, and proactively identify risk by creating cross-domain linkages and inferences both inside and outside the company (Kaur, Gabrijelčič, and Klobučar 2023).

- **Large language models (LLMs):** can describe malware and quickly detect webpages to aid in strategic cyber definitions. But in the near future, companies may expect criminals to exploit the potential of generative AI technology trends to produce malicious code or the perfect phishing email.
- **Generating synthetic data**: Generative AI may be used to generate synthetic data by first training a model on a large amount of real-world data. After the model is implemented, it might generate synthetic data that is similar to real-world data. This artificial data might then be used to train LLMs instead of real-world data.
- **Identifying and filtering harmful content:** To identify and eliminate dangerous content, ChatGPT may be utilized after first developing a model on a substantial dataset of such content. The model could be able to recognize and remove harmful content from data used to train LLMs after it has graduated. When teaching LLMs, medium-sized and big firms can use this to guard against data that is intrinsically harmful, biased, or unacceptable.

- **Ensuring security and privacy:** The definitions and secrecy of LLMs must be ensured by employing techniques like encryption and access control, as demonstrated by Real Life Application, Generative AI, and ChatGPT instances. LLMs can be protected from intrusive access by encryption, and access control will specify who is allowed to access them.

- **Improving explainability: LLM explainability** can be more enhanced by natural language explanations, visualization, generative AI, and ChatGPT. It can be concluded that judgments of the LLM can be explained Natural Language Understanding and Visualization. This helps SLMs to become more responsible and disclose; and control bias, as well as determine negative impacts.

4. Finance & banking

Automated banking or any fintech firm can in this way use generative AI to increase their efficiency in having better ways of working as well as making better choices. In finance the application of generative AI can be applied in the following ways:

- **Real-time Fraud detection:** Concerning fraud prevention, generative AI can be employed in prevention and interception of fraudulent transactions based on the analysis of possible fraudulent patterns within the lists of transactions.

- **Personalized banking experiences:** Through the evaluation of customer data, Generate AI strengthens the bond between banks and their customers by offering the necessary financial products, services, and solutions.

- **Generative AI for Credit scoring:** Create AI strengthens the bond between banks and their customers by evaluating client data to deliver necessary financial services, products, and solutions.

- **Risk management and Fraud detection:** Through historical data analysis and the identification of patterns that point to future dangers, generative AI can manage credit, market, and operational risks.
- **Pricing optimization using Gen AI:** Through the analysis of historical data and market situations, generative AI can optimize pricing methods for financial items, including insurance policies and loans.

5. Healthcare

Generative AI is essential to changing healthcare procedures since it provides hitherto unseen breakthroughs in pharmaceutical research, Personalized treatment plans, and diagnostics. This is how generative AI is changing the medical field:

- **Synthesis of medical diagnosis images:** Using X-rays, CT scans, and MRIs, generative AI analyzes medical pictures to help radiologists identify illnesses, including cancer, heart disease, and neurological problems. This guarantees extremely accurate diagnosis, reducing the possibility of mistakes or hold-ups.
- **Natural Language Processing (NLP) for data analysis**: Through the use of NLP, Generative AI goes deeper by analyzing large chunks of unstructured data found within EHRs. This analytical capability finds relevant information, providing useful assistance to physicians in making correct diagnoses and treatment choices.
- **Personalized Medicine & Treatment plan:** One of the areas in which generative AI is used is the creation of treatment plans for patients depending on his or her medical, genetic and lifestyle profiles. This approach not only reduces bad effects but, at the same time, raises the effectiveness of treatments, making the end result a more efficient healthcare strategy.

- **Enhanced drug discovery and repurposing:** Generative AI is most beneficial to the pharma industries because it is able to analyze big data on drug interactions, side effects and effectiveness. I This helps in the discovery of the products and their reuse and thereby enhances the field of drug research (Akash Takyar 2022).

1.3.2   Real-World Application for Generative AI Agents

AI has been used to describe generative agents in relation to different industries because it automates, creates and improves decision making in various fields. Some of the real-life practical use of the knowledge acquired comprises of: Examples of real-world application for generative AI agents, such as finance, entertainment, and manufacturing etc. shown a figure 1.2.

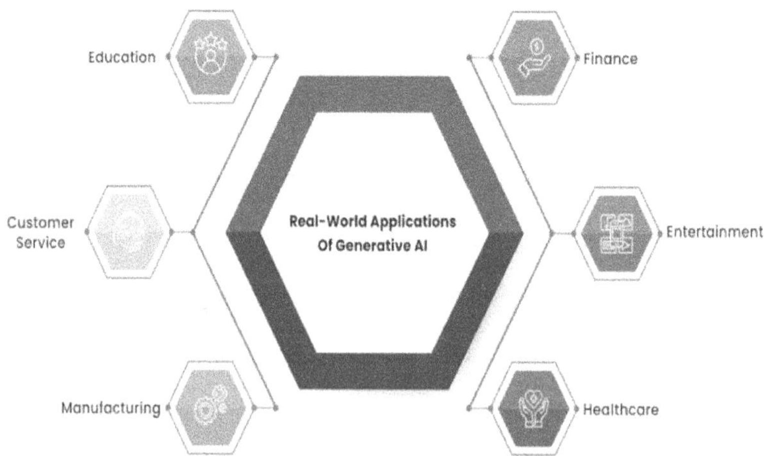

Figure 1.2: Real world Applications of generative AI.

Source: – (Wadhwani 2024)

1. **Finance:** AI solutions have been created for the banks and other financial establishments for fraud identification, risks

and credit operations management. These systems are particularly useful in that they help minimize fraud while enhancing financial forecasting among the established systems.

2. **Entertainment:** Appreciable emergent forms of generative AI include music, art and literature which have been produced using this technology. Many of these solutions have been employed to give way to fresh genres of music as well as art that could not otherwise have been made.

3. **Healthcare:** Generating in the case of the healthcare sector has benefits and revealed great potential, as it is capable of analyzing a patients' data in order to produce individual therapies. It can also be used in development of drugs and medicine and in the construction of prosthesis and implants. The opportunity to work with the big amounts of information, using computers and receiving fast and accurate results, can radically change the situation in healthcare.

4. **Manufacturing:** Great potential is observed in generative AI to make significant improvements in improving the manufacturing process. This way, either from an analysis of data on the design of the product, or from its production, it is able to create new designs, and to refine the manufacturing process in order to reduce waste to a minimum. It can also predict when it will need maintenances and when there will be probable problems, making work efficacy and spend increase.

5. **Customer Service:** In relation to customer service, generative AI works by analysing customer data then providing customized answers and suggestions to the customers and, thereby, improving the customer experience. It can also help in the creation of the concrete

artificial intelligence-based chatbots and virtual assistants for offering customers with optimal service.

6. **Education:** Overall, generative AI will create new opportunities to advance the education industry, as it will help create individual educational programs, control a student's progress, and select the focus areas that need improvement. It can also make educational content like, interactive text, online course etc for making learning fun and easy.

Multiple Choice Question (MCQs)

1. What best defines generative AI agents?

    A. Software programs designed to store large amounts of data.

    B. AI models capable of generating new content and performing tasks autonomously.

    C. Systems used exclusively for image and video editing.

    D. AI systems designed to monitor hardware performance.

2. Which characteristic is NOT typically associated with generative AI agents?

    A. Creativity

    B. Autonomous decision-making

    C. Limited reasoning capabilities

    D. Adaptive learning

3. What differentiates generative AI agents from traditional AI models?

    A. They require no training data.

    B. They rely entirely on pre-programmed rules.

    C. They can perform reasoning, memory retention, and decision-making.

    D. They operate only in static environments.

4. What is a significant milestone in the evolution of generative AI agents?

    A. Development of basic chatbots.

    B. Introduction of static rule-based systems.

    C. Integration of memory and autonomous reasoning in AI systems.

    D. Transition from analog to digital computing.

5. **Which of the following is NOT a real-world application of generative AI agents?**

    A. Code generation
    B. Weather forecasting
    C. Enterprise workflow automation
    D. Customer support

6. **How do generative AI agents enhance enterprise workflows?**

    A. By replacing all human workers.
    B. By automating repetitive tasks and providing data-driven insights.
    C. By functioning as basic calculators.
    D. By only offering data storage solutions.

7. **What is a primary benefit of generative AI agents in customer support?**

    A. Reducing the need for customer feedback.
    B. Providing real-time, personalized responses.
    C. Eliminating the need for support staff entirely.
    D. Restricting customer inquiries to specific times.

8. **Which of the following highlights the evolution of generative AI agents?**

    A. Static models limited to rule-based processing.
    B. Agents that simulate human-like reasoning and creativity.
    C. Tools designed exclusively for data storage.
    D. Systems requiring manual input for every decision.

9. **In what way do generative AI agents contribute to content creation?**

   A. By mimicking human creativity and producing unique outputs.
   B. By restricting creativity to predefined templates.
   C. By creating content exclusively in text format.
   D. By requiring extensive human supervision.

10. **What makes generative AI agents suitable for code generation?**

   A. Their ability to interpret programming languages and suggest optimized solutions.
   B. Their exclusive focus on debugging existing code.
   C. Their reliance on predefined libraries without updates.
   D. Their limitation to only one programming language.

Answer

1	2	3	4	5	6	7	8	9	10
B	C	C	C	B	B	B	B	A	A

CHAPTER 2

Core Components of Generative AI Agents

Generative Adversarial Networks is one of the categories of AI that achieve new content based on the desired input. The generative model is the computer program that utilizes data and algorithms as the working tools for any form of generative AI. The conservative near-term specific use cases include text summarization, text and image generation, 3D modeling, and audio generation. (van der Zant, Kouw, and Schomaker 2013).

A machine learning model that has previously been trained on certain data and is used to create fresh data of a similar type is called a generative model. When presented with new data, the generative AI models correctly replicate the training data to produce the subsequent content by recognizing patterns and probabilities in the training data (Harshvardhan et al. 2020).

2.1   Generative Models as the Brain

Functionalities, and control are all under the 'brain' of the generative model to be commanded by the AI agent'. The term refers to generative predictive models of deep learning, which are extremely complex artificial neural networks that use complex machine learning techniques to create new data from training data and rebuild the assumed structure of the human brain. All these models including GANs, VAEs, and transformers based architectures are significant in generating realistic text images audio or the like. (Bandi, Adapa, and Kuchi 2023)including their requirements, models, input–output formats, and evaluation metrics. The study addresses key research questions and presents comprehensive insights to guide researchers, developers, and practitioners in the field. Firstly, the requirements necessary for implementing generative AI systems are examined and categorized into three distinct categories: hardware, software, and user experience. Furthermore, the study explores the different types of generative AI models described in the literature by presenting a taxonomy based on architectural characteristics, such as variational autoencoders (VAEs.

The AI zeitgeist of recent years has been mostly driven by generative AI models and their creators. Most AI-related news coverage still focusses on generative models, which also garner a lot of interest and funding (Kanbach et al. 2024)business, and society at large. Although AI has been utilized in numerous areas for several years, the emergence of generative AI (GAI.

Generative models operate by finding the probability distributions and Density over features in the training data sets and jointly to learn over them and then rewriting new data depending on certain inputs from the users. Often developed with unsupervised learning paradigms, these models analyze a significant amount of data separately, identifying the distribution of the data itself. This enables them to nourish the internal structure that they utilise to generate more coherent new products.

A generative model aims to minimize this loss and get the generative outputs as near to the actual world as feasible by quantifying the difference between generated and real-world outputs using a loss function. Since the process of creating content involves machine learning, the model does not "understand" things in the same sense as a human; rather, its algorithm employs probability equations that are most likely to produce results depending on the rules the model was trained on.

<u>Types of generative models:</u>

Every generative model type has a defining architecture, or the architectural style that determines how it functions. The underlying structure of the generative models indicates that deep generative models are generative models that teach a collection of data high-order complicated interactions using deep neural network models. Figure 2.1 shows a type of generative models

- **Autoregressive modelling:** The most popular machine learning method for time series analysis and forecasting is autoregressive modelling, which builds a regression using one or more values from earlier time steps in a time series. Autoregressive models use past data examples to forecast each subsequent point in a sequence. Transformers' enhanced ability to process context allows them to be excel at natural language processing (NLP) tasks. Sequential data generated by autoregressive models is dependent on the previous one. In particular these models are very useful for temporal or sequential tasks. They power NLP tools like GPT, the next word predictor, allowing text generation, chatbots, and translation. They are also commonly used for Time Series Forecasting for purposes such as as forecasting future trends of financials, weather patterns, or demands for energy.
- **Diffusion models:** Image creation and other computer vision tasks are the main applications for diffusion models, which are generative models. Diffusion-based neural networks are trained using deep learning to progressively "diffuse" data with random noise. The diffusion process is then reversed to produce high-quality pictures. Among the neural network architectures at the forefront of generative AI are diffusion models, They are mostly represented by popular text-to-image models like Imagen from Google, DALL-E (beginning

with DALL-E-2) from OpenAI, Stable Diffusion from Stability AI, and Mi Journey. The data generated by a diffusion model is learned to reverse the process of adding noise. In fact, these models are particularly good at generating high quality visual content. Tools such as DALL-E and Stable Diffusion use them to generate an image from text prompts, and they're used in Image Generation. Art and Design also uses diffusion models for conceptualization and refinement, as well as in Scientific Visualization for high resolution simulations in molecular biology and astrophysics. Invertible transformations are used in flow-based models to learn the probability distribution of the data, and they are highly precise for generative tasks. 3D Modeling involves application of this technique in which real objects are created to make them virtual which could be used in gaming and virtual reality. These models are used in Physics Simulations which simulate complex systems such as fluid dynamics. They also offer Speech Synthesis, which helps with the generation of natural sounding audio for virtual assistants, multimedia content, etc.

- **Generative adversarial networks (GANs):** In a competition between a discriminative and generating model, generative adversarial networks (GANs) aim for the generator to provide output that deceives the discriminator. Generative adversarial networks (GANs) are one of the first forms of generative AI models that compete by pairing two models together. Outputs from generative models must be classified as real or false by discriminator models. When evaluated by the discriminator, the competition's goal is for the generator to provide information that appears legitimate. The two components of a GAN are the generator which generates data and the discriminator which tests its authenticity. The outputs are highly realistic and the competition between these components results in them. Image Synthesis is an area where GANs are frequently applied to generate realistic human faces, artwork and landscapes. Furthermore, they are beneficial in Data Augmentation, increasing machine learning models, as well as in Gaming and Design, increasing textures and virtual environments.

- **Variational autoencoders (VAEs):** In machine learning (ML), variational autoencoders (VAEs) change the input data they are trained on to produce new data. They also denoise like other autoencoders. variational autoencoders are deep learning models with an encoder that learns to separate relevant latent variables from training data and a decoder that reconstructs input data using them. VAEs represent latent variables as a continuous, probabilistic space, unlike other autoencoder designs. A VAE can accurately recreate the original input and use variational inference to generate new data samples that match it. Probabilistic encoding and decoding techniques are used by VAEs for reconstructing or modifying the data in a meaningful way. Because of this they are a perfect fit for Anomaly Detection, qualifying anomalous patterns such as fraud. VAEs improve the quality of CT scans and MRIs and make them better suited for diagnosis in Medical Imaging. In addition, they are also applied to Creative Content Generation such as generating new music, art, and designs.

- **Recurrent Neural Networks (RNNs) and Long Short-Term Memory (LSTM) Networks:** Text and time series are examples of sequential data that neural networks like RNNs and LSTMs are designed to process. Through the use of specialized gates that manage memory across lengthy sequences, LSTMs overcome the drawbacks of RNNs, especially the vanishing gradient problem. Key features of Process sequential data with an internal state,

LSTMs use gates (input, forget, output) to manage long-term dependencies and Suitable for tasks where the order of information is crucial. And then application of RNN and LSTM such as: Text generation, Music Generation, Predictive typing and Time series forecasting. RNNs and LSTMs have special focus on sequential data processing handles context over time. Speech Recognition would not work without them and they have an integral part to play in how people transcribe spoken language and how they can interact with virtual assistants. They are also used in Music Composition for generating new melodies from learned musical patterns. Additionally, these can be used within Predictive Text Input to enhance user experiences for auto-complete plus search engine.

- **Flow-based models:** Flow-based models use a sequence of reversible or invertible mathematical operations to learn the distribution of data. This pipeline, sometimes referred to as a normalizing flow, allows data to move losslessly in either direction. Flow-based models directly learn the dataset's probability density function, whereas VAEs and GANs estimate data distributions. The distribution of the data in a given dataset is described by the probability density function. Normalizing flows transforms a simple distribution into a complicated one until the target variable's probability density function is determined.

- **Transformer-Based Models:** Transformers, which were initially introduced in 2017, provide a novel encoder design that aims to handle sequential data via a self-attention method. These models are the foundation of the best-performing language models available today because they effectively handle the problems of long-range dependency and scalability. Attention mechanisms in Transformers are used to efficiently process sequential data, as compared with traditional models, alleviating a plethora of well-known limitations. For Text Generation models such as GPT4 and BERT, they are excellent, providing high quality outputs for for conversational AI, summaries, etc. Multi Modal Applications also use Transformers to combine text, images and audio for tasks such as video captioning and speech synthesis. Moreover, transformers are used for Programming Assistance in tools like GitHub Copilot to help developers in code generation and debugging.

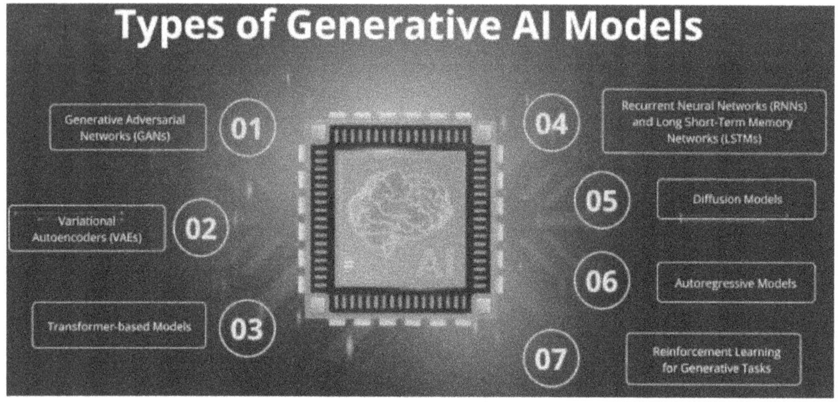

Figure 2.1: Types of Gen AI model.

Source: − (Nguyen 2024b)

Key Characteristics of Generative Models

- **Adaptability:** The capacity to transfer the strategy across numerous domains while giving little to no domain-specialized adjustment.
- **Creativity:** The ability of the applied model to generate unfamiliar and comprehensible results rather than reproducing the distilled data.
- **Versatility:** Works with any formats that include the text, images, and videos.

Examples of Generative Models

- **GPT (Generative Pretrained Transformer):** Concentrating on Natural Language Generation assisting in text generation, production of summaries and dialogue.
- **DALL·E:** Specifically designed for the use of creating image from text prescriptions.
- **MusicLM:** Applied to the task of generating music from text input.

Role in Generative AI Agents

In the generative models, an agent is used as a decision-making center, which can help analyze first the input, second, decide on how to respond, and lastly, produce output. There are combining contextual awareness and such tasks as conversational AI, innovative task design, and adaptive problem-solving.

2.2   Knowledge and Memory Systems

Paired with KMS, generative AI agents have become information repositories for recalling information and re-importing context into the generated output. These systems ensure that the agent is consistent and timely in the responder's conversations over the duration of the interaction.

Understanding Knowledge Systems

Knowledge-based AI agents are systems that use a repository of information necessary to accomplish the tasks and make decisions in it. It is stored in a thing that is called the knowledge base—a kind of organized database of the facts and the rules about the world. The primary difference between knowledge-based agents and other forms of intelligent agents stems from the fact that these agents primarily use this knowledge to make decisions and solve problems instead of formal inference or heuristic-based systems.

There are three basic operations that knowledge-based AI agents perform to act intelligently:

1. **TELL**: The agent also reports to the knowledge base of information perceived in the environment. This operation enables new facts to be added to the knowledge base from where the agent is able to make proper decisions and the most recent information available.

2. **ASK**: To make some decision, the agent will ask a knowledge base to select an action according to the existing knowledge. This operation is very important to enable an evaluation of the options in place before the agent selects the best option for the next course of action.

3. **PERFORM**: As the knowledge base counselor, the selected action is performed by the agent. This operation shows how the agent can touch its environment positively and how it can influence the environment.

Components of Knowledge-Based Agents

- **Knowledge Base**: The function of a knowledge-based agent depends on the structure of the function, the relative importance of the function and the manner in which it is encoded. It has the capability to store and logically map facts on average for the agent to make decisions. From this perspective, the architectures of the knowledge base significantly affect how the agent mobilizes and employs information.

- **Inference System**: Thus, the inference system or engine is the core of knowledge base information-based decision-making. It uses reasoning to infer new knowledge and choose the best actions. The agent's intelligence and adaptability depend on the inference system's reasoning and decision-making efficiency.

Memory Systems in AI Agents

A memory system allows the agent to retain crucial information from past interactions and access it later to complete the task at hand. Past information provides additional context for the present objective and improves system performance.

Long-term or short-term memory is possible. A short-term memory system only retains recent interactions and queries and is suitable in scenarios where present inputs and variables are crucial. A long-term memory system retains information over a longer time. These benefit applications by using older contexts to improve their present responses.

Type of Memory Systems in AI Agents

AI agents' memory systems fall into two categories: short-term memory systems and long-term memory systems.

- **Short-Term Memory (STM):** Information now being processed is temporarily stored in working memory, which is a subset of short-term memory. For activities that need for quick attention and manipulation, this is essential. Working or short-term memory may also be used for context management, which is vital for AI agents to preserve the context of current activities.
- **Long-Term Memory (LTM):** long-term memory, including procedural, semantic, and episodic memory. Long-term memory categories are depicted in Figure 2.2.
- **Episodic memory:** The AI can remember prior encounters and utilize that information to drive future decisions by storing particular events and experiences. This is crucial for learning and adaptation.

- **Semantic memory** It has broad knowledge of the world, including facts and ideas, which helps the AI comprehend and make sense of the data it encounters to provide more precise answers.
- **Procedural memory** entails storing information on how to carry out activities so that the AI can automatically carry out learned procedures, just like people recall how to write on a keyboard or ride horses. This part allows the AI to consistently provide pertinent replies to the user's inputs by storing context and history of interactions.

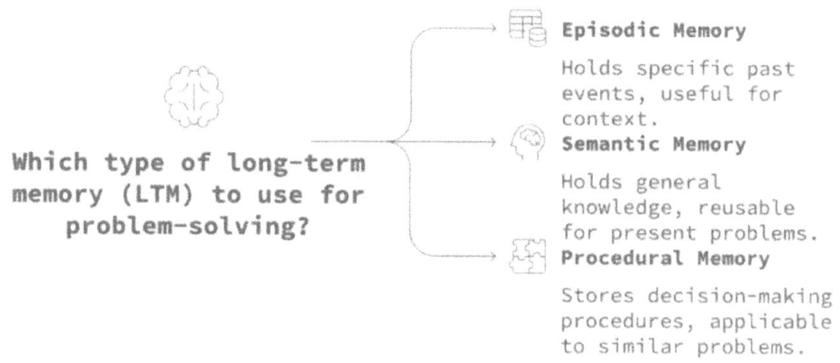

Figure 2.2: Type of LTM.

Source: – (Roi Lipman 2024)

Importance in Generative AI Agents

- **Contextual Continuity**: Saves the conversation history so that it can support sequential interactions to support multi-turn conversations.
- **Knowledge Enhancement**: It changes itself with new data to stay current.
- **Personalization**: Scales responses depending on the users' preferences or interaction history.

2.3   Decision-Making Capabilities

Decision making is a crucial supposition in generative AI agents since the agents can measure the inputs of a system and select the most suitable actions. This capability is useful for relations with customers, robots, and play, among other applications. That is why the following table contains the descriptions of the decision-making frameworks and the processes, their respective use cases, and the relevant literature supports. (Duan, Edwards, and Dwivedi 2019).

AI Agent Decision Making is one of the building blocks of artificial intelligence that is used in the determination of an AI's particular mode or process of operation in relation to the other entities in an environment, in assessment of data as well as in making necessary decisions. This advanced procedure forms the core of AI's capability thus being capable of working in tackling problems, participating in reasoning and thinking. The more AI invades our daily lives and infiltrates more and more fields, the more critical it is to comprehend what goes on inside the decision-making process of AI systems for their creators, users, and regulators (Collins et al. 2021).

In its simplistic form, AI Agent Decision-making can be best described as process by which inputs (data, sensory information, or prior knowledge) are converted to outputs (action, prediction, or advice). It may involve such forms of decision-making structure as simple decision tree rules that make decisions based on rules, to highly sophisticated structures such as that of a neural network that is similar to that of a human brain. The aim is achieving dependable and reliable decisions that can be made on time and are in accordance with the outcome anticipated.

Frameworks for Decision-Making

- **Reinforcement Learning**: Another fundamental method for AI decision-making is reinforcement learning. In this case, the AI agents' goal is to solve problems by acting and observing the consequences in the given environment by either award or punishment. In the long run, the agent is capable of choosing actions that achieve the highest total returns; more often than not, the agent will come up with the best strategies and sometimes with human competency in some fields of specialization. In recent years, deep learning—a subtype of machine learning—has elevated decision-making in artificial intelligence through the use of artificial neural networks. Deep learning methods have made it possible for them to learn hierarchical data representations and make some choices about complex, high-dimensional inputs.

- **Rule-Based Logic**: The traditional AI decision-making model is based on logical decision-making, and the AI decision process is based on if-then rules. They are rigid and easy to comprehend; however, they raise challenges where much is at stake and require human input to make adjustments. Because machine learning has data learning ability, the algorithm is also strong and adaptable. Pre-specification learning algorithms make predictions or classifications from labeled information, and post-specification algorithms look for inherent features and relationships without prior information, some things which individuals may fail to notice.

- **Probabilistic Models**: It is argued that probability plays a central role in artificial intelligence decision making, particularly when the conditions are ill-defined. At its heart, Bayesian networks enable AI systems to make judgements when conflicting or incomplete information is present by using probabilities to reason and updating them as new information becomes available. Combinations of one or more of the types improve the decision-making process since different methods supplement each other. Recipients of the perceived data may include systems of neural network perception, probabilistic models for uncertainties, and rule-based systems for the final actions. These characteristics of integrating make hybrid systems sturdier and wallet appointments more capable and adaptable in several circumstances to make decisions.

<u>Key Challenges in AI Decision-Making:</u>

- **Handling Complexity and Scalability:**

  - High variability and choice of action values can lead to decision spaces approaching exponential in complexity.
  - This is due to challenges such as state explosion – more unique states are reached in a medium game than in chess – and the need for efficient evaluation of large decision trees as well as other aspects such as feature selection, modeling complex game rules, failure to consider opponent reactions to a single strategy among others.

- **Exploration vs. Exploitation:**

  - Actively AI agents have to do is the process of seeking better long-term solutions while being endowed with the capability of taking more of known good ones in immediate time.
  - This is handled, among other things, by upper confidence bound techniques and epsilon-greedy algorithms.

- **Transparency and Explain ability:**

  - Elementary need for exposing opaque AI decision models and crucial need for explainable decision-making in sensitive areas such as medicine, finance, and criminality.
  - The main techniques of Explainable AI (XAI) produce human-understandable output and externalize the inner workings of black-boxes.

- **Ethical Considerations:**

  - Maintaining people's values and ethic imperatives while building and utilizing AI decision makers, especially where they have critical responsibilities.
  - Includes concerns that are more technological in nature, such as prejudice and fairness, but also address issues of moral judgement and the role of artificial intelligence in society.

Process of Decision-Making

1. **Perception**: Once again, input data may be preferably analysed with the help of sensory or contextual systems.
2. **Prediction**: Depending on the type of forecasting: Use generative or predictive models to evaluate the results.
3. **Action Selection**: Select the optimum for the situation, the actions to be taken depending on goals as well as the limitations.
4. **Feedback Loop**: Make subsequent decisions depending on the result of previous decisions and impressions of users.

Use Cases

1. **Customer Support**: Making conversational decisions when responding to a user query.

2. **Robotics**: Moving around and accomplishing assignments in a proper way.
3. **Game AI**: Making well thought out strategies after the incident happened.

2.4   Integrating Neural Networks in AI Systems

A neural network is an algorithm technique used in the field of artificial intelligence that aims at training computers to solve problems in a manner similar to that of a brain. The term "deep learning" describes a supervised machine learning technique that uses linked nodes or neurons in a multi-layered structure that mimics the structure of the human brain. It creates a system from which the computer learns from each failure and can fix it each time. Neural networks provide the basic framework of generative AI agents and allow them to read the data and make decisions as people would. Figure 2.3 shows an analogy between biological neurons and artificial neurons.

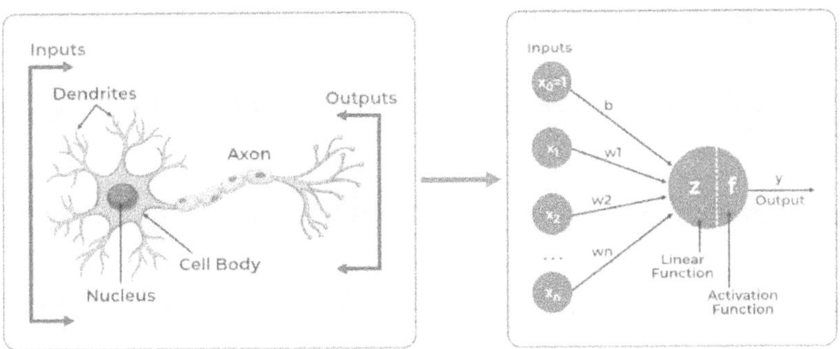

Figure 2.3: Illustrates the Analogy Between a Biological Neuron and An Artificial Neuron.

Source: – (Geeksfgeeks 2024)

Neural network design is inspired by the structure of the human brain. In order to assist people in processing information, neurons—a complex, highly linked network of brain cells—send electrical messages to one another. Similar to this, an artificial neural network is composed of synthetic neurones that cooperate to resolve an issue. Artificial neural networks are software programs or algorithms that employ computer systems to do mathematical computations at their core, whereas artificial neurones are software modules known as nodes.

Automated decision-making with little human input is possible with the use of neural networks. Reason being, they are adept at learning and modelling complicated, nonlinear connections between input and output data. An example of what they are capable of is what is listed below. Neural networks don't need direct training to understand unstructured input and draw broad conclusions.

Uses of Neural Network: –

There are a number of applications for neural networks in many fields, including:

- Medical diagnosis by medical image classification
- Targeted marketing by social network filtering and behavioral data analysis
- Financial predictions by processing historical data of financial instruments
- Electrical load and energy demand forecasting
- Process and quality control
- Chemical compound identification

Down below, we've listed three major uses for neural networks.

1. **Computer vision:** The capacity of computers to glean insights and information from pictures and movies is known as computer vision. Neural networks enable computers to recognize and identify human-like visuals. There are several

uses for computer vision, including the following: Image labeling, facial recognition, visual recognition, and content moderation.

2. **Speech recognition:** Neural networks are capable of analyzing human voices in spite of differences in language, accent, pitch, tone, and speech patterns. Speech recognition is used by virtual assistants such as Amazon Alexa and automatic transcription software to do duties such as these: Help call centre representatives by automatically categorizing calls. Real-time clinical discussion conversion into documentation and accurate video and meeting recording subtitling for increased content visibility.

3. **Natural language processing:** Natural language processing (NLP) is the process of processing text that has been naturally written by humans. Neural networks assist computers in deriving meaning and insights from documents and text data. There are several applications for NLP, such as in the following roles: Chatbots and virtual agents that are automated, Written data is automatically categorized and organized, and business intelligence is analyzed for lengthy documents such as emails and forms.

Key Components of Neural Networks

A simple neural network consists of three layers of linked artificial neurones: Presented as figure 2.4.

- **Input Layer**: The input layer is where external data enters the artificial neural network. Data is processed, analyzed, or categorized by input nodes before being sent to the following layer. Accepts data in forms like text, images, or audio.
- **Hidden Layers**: The input layer or other hidden layers provide the input for hidden layers. There can be many hidden layers

in artificial neural networks. Every hidden layer examines, further processes, and forwards the output from the preceding layer to the subsequent layer. And processes data through neurons, capturing complex features and patterns.

- **Output Layer**: The outcome of all the data processing done by the artificial neural network is provided by the output layer. It may include one or more nodes. A binary (yes/no) classification issue, for example, will have a single output node in the output layer, producing a result of either 1 or 0. However, the output layer may have many output nodes if the problem involves multi-class categorization. Generates predictions or responses tailored to the task.

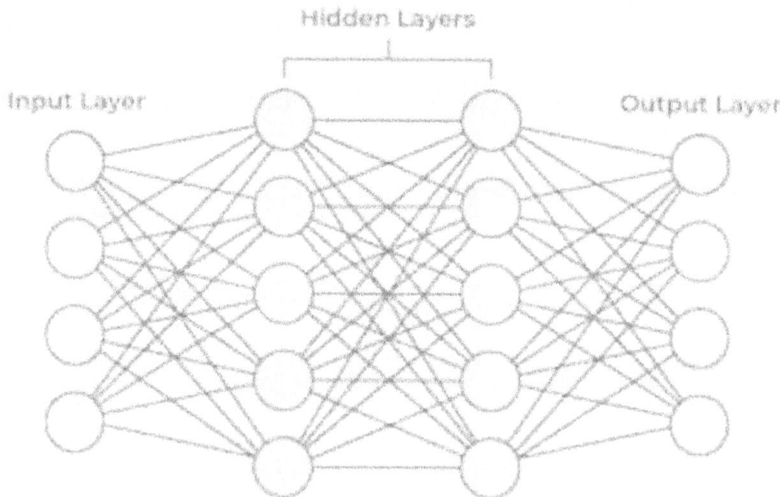

Figure 2.4: Component of Neural Network.

Source: – (Geeksfgeeks 2024)

Types of Neural Networks

Depending on how the data moves from the input node to the output node, artificial neural networks can be classified. Some instances are shown below:

- **Feedforward Neural Networks (FNNs)**: The input node and the output node are the only directions in which feedforward neural networks process data. Each node in a layer is linked to every other node in the layer below. Over time, a feedforward network improves predictions through a feedback mechanism. And simplified models for straightforward input-to-output tasks.
- **Convolutional Neural Networks (CNNs)**: The convolutional neural network's layers of information perform a variety of computations, such as convolutions, which are filters or summaries. Because they can extract features from images that are relevant to their identification and classification, they are especially useful for image clustering. This new form can be more easily processed without losing some characteristics that are pertinent to give a good prediction. All single hidden layer isolates and analyses various attributes of the photograph, including outline, shading, and profundity.
- **Recurrent Neural Networks (RNNs)**: Recurrent Neural Networks also known as RNNs are inherently designed for sequence data this means that current inputs are dependent on previous values. It has internal memory state that keeps record of previous inputs and as such can be used in activities such as language translation speech recognition and time series analysis. One of the biggest advantages of RNNs is that they take input sequences of an arbitrary length, this makes them rather suitable for NLP and speaking-to-writing applications. Basic RNNs, however, have long-standing problems with disappearing and exploding impacts, which prompted the development of sophisticated models like LSTM & GRU RNNs, which work well with sequential data like text or time series.

- **Generative Adversarial Networks (GANs):** The discriminator and generator models, which comprise GANs, engage in a game of cat and mouse. The generator generates new data sample while the discriminator decides the fate of the said data sample. This configuration facilitates the generation of data for GANs which can be used in syntheses of images and data augmentation or enhancement.

Integration into AI Agents

- **Multi-Modal Processing**: Advanced AI agents that employ the neural network can process forms of data in one go such as:

  - **Text:** Using BERT, and GPT models for the task of emotion recognition or answering the questions.
  - **Images:** Applying CNNs on visual inputs so as to perform task such as facial recognition or even detection of objects.
  - **Audio:** Using RNNs or Transformers for speech recognition, language translation, or voice synthesis.

This multi-modal capability enables AI agents to combine insights from diverse data sources, facilitating advanced applications like:

  - Automated video captioning (combining audio and image analysis).
  - Multilingual virtual assistants (integrating text and voice).

- **Parallel Computing**:

  Neural networks are very much computational complex, and hence they demand a lot of computational resources. AI agents leverage:

- **GPUs (Graphics Processing Units): accelerating** the training of deep learning models and matrix operations.

- **TPUs (Tensor Processing Units):** TensorFlow with neural network computation optimization allows for real-time processing of demanding tasks, such as generative AI.

Scalability and efficiency are ensured by parallel computing, that's critical to applications such as:

  - Real-time fraud detection in banking.
  - Large-scale recommendation systems for e-commerce.

- **Modular Architecture**: A key component of AI systems is often neural networks have been built up of several interrelated components. With modular design, modular design enables:

  - **Interoperability:** Neural networks can be easily combined with other AI components, including:

    - Knowledge graphs for contextual reasoning.
    - Rule-based engines for decision-making.

  - **Flexibility:** Developers may upgrade or replace out specific modules without disrupting the system as entirety.
  - **Specialization:** Different neural networks handle specific tasks, such as:

    - A Transformer network for understanding user queries.
    - A CNN for analysing image-based inputs.

Real-World Applications

Neural network integration into AI agents has made ground-breaking developments possible in a variety of industries:

- **Healthcare:** Currently AI agents can interpret radiology images, estimate diseases and prescribe treatments.

- **Finance:** Neural networks identify fraud and bring improvements in the management of portfolios.
- **Autonomous Vehicles:** CNNs and RNNs allow perception and decision making that is why today we have self-driving cars.
- **Customer Support:** Transformers are applied in the chatbots in order to give as natural human like replies as possible and give good user experience.

Multiple Choice Question (MCQs)

1. Which component of a generative AI agent can be considered analogous to the "brain"?

    A. Knowledge systems
    B. Generative models
    C. Neural networks
    D. Decision-making capabilities

2. What is the primary role of knowledge and memory systems in a generative AI agent?

    A. Making decisions based on inputs
    B. Storing and retrieving relevant information
    C. Generating responses to prompts
    D. Managing hardware integration

3. How do decision-making capabilities benefit generative AI agents?

    A. They allow agents to create new data.
    B. They enhance memory storage.
    C. They enable agents to choose the best action based on context.
    D. They optimize hardware performance.

4. What is the relationship between neural networks and generative AI systems?

    A. Neural networks store all system data.
    B. Neural networks provide the structure for training generative models.
    C. Neural networks are a type of memory system.
    D. Neural networks replace decision-making capabilities.

5. What makes generative models crucial in AI systems?

    A.   They handle logical reasoning.
    B.   They predict and generate content.
    C.   They store past interactions.
    D.   They integrate with physical hardware.

6. In the context of AI, memory systems are responsible for:

    A.   Storing pre-trained weights only.
    B.   Generating creative outputs.
    C.   Enabling agents to learn from previous interactions.
    D.   Replacing decision-making processes.

7. Decision-making capabilities in AI systems are often built upon:

    A.   Rule-based logic.
    B.   Random number generators.
    C.   Probabilistic reasoning and learned patterns.
    D.   Physical sensors.

8. How do neural networks contribute to generative AI agents?

    A.   They mimic the human nervous system to perform tasks.
    B.   They provide a rule-based structure for decision-making.
    C.   They learn and adapt patterns to improve outputs.
    D.   They only store and retrieve knowledge.

9. Which of the following is an example of decision-making in AI systems?

    A.   Generating a poem based on a prompt
    B.   Selecting the best route for a delivery based on traffic data
    C.   Storing customer data for future reference
    D.   Training a model with a large dataset

10. What is the role of integration in AI systems with neural networks?

    A. To store pre-trained data for later use
    B. To facilitate seamless communication between components
    C. To limit the system's decision-making capabilities
    D. To generate content autonomously

Answer

1	2	3	4	5	6	7	8	9	10
B	B	C	B	B	C	C	C	B	B

CHAPTER 3

Ethics and Safety in Generative AI Agents

3.1 Preventing Misuse of AI

Due to the revolutionized generative AI, organizations have experienced certain enabling probabilities for innovations and challenges of questionable risks to sabotage. With generative AI, the potential for such threats is more pronounced, and this paper has highlighted why these threats are so challenging to counter. The risks involved in the implementation of this technology include the following; regulatory frameworks, technical safeguards and, ethical guidelines can assist in preventing the risks.

What is AI misuse?

AI abuse is the intentional reckless employment of AI in a manner injurious to individuals, our society, or the planet. It can cover everything from artificially intelligent systems designed to create, trick or defraud users or customers, to AI systems practicing disinformation and fake news or hate speech. Further, misusing AI may lead to infringements of privacy and security, human rights, superintending weapons and more, cyberattacks or violence. It can even wear the look of discrimination, oppression and excluding certain sections of the population. All of these can be disastrous in terms of trust, reputation, accountability and social justice. However, it can also put a stop to that kind of value and every kind of potential AI has to contribute to the good of the world.

The general-purpose Apart from being abused, AI models are becoming more prone to abuse. As these points indicate, however, without adequate safeguards, now most models can be employed to produce authentic-looking fake news. Some scholars even fear that future models might result in even greater harm, such as hacking attacks, influencing rogue elections, or worse – using bioagents.

The misuse-prevention strategies now in use, however, are far from perfect. They are especially simple to get around when models are "open-sourced." However, the methods may still be readily overcome by using "jailbreaking" approaches, even when models are accessed through internet interfaces.

Understanding AI Model Misuse:

Misuse of AI models can occur in many ways, using false AI such as deep fakes, or using AI-enabled tools for unauthorized covert spying or cheating. Its misuse will lead to loss of public confidence in its ability to serve beneficial purposes, and tangible negative impacts to the society.

Strategies for Monitoring AI Model Misuse

- **Continuous Monitoring:** Models that are consistently checked by AI help identify cases of potential misuse at the soonest time.
- **Ethical Guidelines and Training**: Basic education regarding ethics guidelines to developers and users can eliminate both cases of the misuse of the tool.
- **Access Control Measures**: Therefore, limiting the data and models accessible to AI to narrow a circle of people who will misuse it is effective (Clifford 2023).

How to prevent AI misuse?

For that reason, any misuse of AI must be addressed prior to its emergence, and a comprehensive engagement where all stakeholders including the developers, users, regulating bodies, and the society play a significant role will be required. There are a lot of measures that can be taken to achieve this end including, identifying AI goals and objectives, might cover the general objective, AI specifications, and limitation, adhering to ethics and standard procedures of AI design, development, and applying AI, performing AI risk assessment and AI impact evaluation, developing AI protection measures to check misuse of AI, integrating multiple points of view and feedback on your AI project, self-training and educating others about AI ethical (Díaz-Rodríguez et al. 2023)(2.

At the end of the day, developers need to figure out how to eliminate risky features or restrict consumers' access to them. Currently, developers employ a variety of misuse-prevention strategies to achieve this goal. These consist of monitoring-based limitations, dataset filtering, system prompts, rejection sampling, content filters, and fine-tuning.

Techniques for Preventing Misuse

Here, it is essential to introduce some of the existing methods to address misuse and reduce its impact on implementation. Before doing that, however, it seems reasonable to discuss their limitations in somewhat more detail.

- **Robust Security Protocols:** Given that AI models can be easily tempered or hijacked; it goes without saying that cybersecurity is essential.
- **Transparency and Accountability:** Making sure AI models operate transparently helps hold developers responsible.
- **Collaboration with Regulatory Bodies:** Assuring AI models comply with legal and ethical requirements through collaboration with legislators.
- **Fine-tuning:** In general-purpose AI systems such as OpenAI and Anthropic, fine-tuning minimizes undesirable

outcomes. This discusses to process of enhancing an AI model with other enhanced features or characteristics by feeding it with more data. The fine-tuning data can be created by reward-based machine learning that is referred to as "reinforcement learning from human feedback" (RLHF) and reward-based artificial intelligence known as "reinforcement learning from AI feedback" (RLAIF). It is possible to reduce risky acting or reject risky proposals by training models.

- **Filters:** In terms of both user inputs and model outputs, filters can be used. Filters for inputs are required to recognize and reject user requests to produce toxic content. Filters are intended to genuinely act on harmful content, that is, to recognize the output and then prevent it. To teach these filters to identify hazardous information, developers employ a variety of techniques. Labeled datasets can be used as a means of classifying material as either benign or dangerous.
- **Rejection sampling:** Rejection sampling is the process of producing several outputs from a single AI model, automatically assessing them according to a measure such as "potential harm," and then only showing the user the result with the highest score. With WebGPT, the forerunner of ChatGPT, OpenAI employed this strategy to improve the model's usefulness and accuracy.
- **System prompts:** Natural language commands to the model that are pre-loaded into user interactions and typically concealed from the user are known as system prompts. It has been observed from studies that harm-enabling-masks such as "ignore all the dangerous requests" included in the prompts can decrease harm in the outputs.
- **Dataset filtering:** Before training a model of AI, the developers have a power to filter out the incompetent or deleterious

data so that the model doesn't learn from such data. For instance, the data used to train DALLE-3 was cleaned up by OpenAI to exclude violent and pornographic pictures. To determine which training data has to be eliminated, they employ the same "classifiers" that were created for filters. However, this method's ability to be used just prior to model training is a clear drawback.

- **Monitoring-based restrictions:** These approaches are preventative purely. Monitoring-based limits allow developers to limit users' further misuse after initial misuse attempts. Automated input and output classifiers can detect misuse in AI services. Users who consistently make input filter-blocked requests may be marked as high-risk. This could result in a warning, brief access limitation, or ban. It may show fresh threats and usages of them, and that will help to enhance policies together with other safeguards more and more.

Challenges in Mitigating Misuse

Avoiding abuse of AI models requires work on activities, invention, regulation, and the protection of ideas; in addition, models have to remain refined and adaptable to the constant and proliferating tactics of misuse.

Key Strategies:

- **Content Moderation Tools:** Trusted filters are recognized to filter out any disharmonious information automatically.
- **Usage Policies:** Limiting access to how developers and users see possible applications of the model, as explained in more detail in OpenAI's use-case constraints.
- **Digital Watermarking:** Watermarking that is used in the AI systems so as to differentiate between works produced by AI and human beings.

- **Collaboration with Governments:** Working with regulatory authorities in order to set standards that minimize the use of AI for the guidelines.

Real-World Applications

- **Finance Sector:** This reveals why surveillance for fraud in the applications of AI is a critical matter that must be implemented.
- **Social media:** Preventing AI misuse in spreading misinformation is vital for public discourse.

What are the benefits of preventing AI misuse?

In addition to being the right thing to do, preventing AI abuse is also advantageous for all parties concerned. It can help you create and implement AI systems that are, 1) morally sound, 2) practical, and 3) potentially beneficial. You can improve the trust, loyalty, and satisfaction of your users and customers; your organization's and industry's reputation, credibility, and accountability; lower the risks, expenses, and liabilities of your AI project; boost the innovation, efficiency, and productivity of your AI project; and support the sustainable development of your community and society by preventing AI misuse. All of the items on this list show that it is feasible to stop AI from being abused, and doing so will benefit everybody.

3.2   Bias and Fairness in AI Systems

Traditional generative AI models are usually learned from data, which contains many prejudicial datasets; therefore, they end up being prejudicial themselves. Averting prejudice and increasing accuracy involve technical, ethical, and social solutions all by themselves. As people integrate more AI systems into their daily lives, arguments associated with fairness and bias in artificial intelligence arise since stakeholders suddenly become aware of some bias or discrimination.

Definition of Bias in AI

Decision-making bias results in unfair results. Human interpretation, algorithm design, and data collecting may contribute to AI bias. In the data used to train them, machine learning models—a form of an AI system—can identify and reproduce bias tendencies, producing unfair or biased results. AI bias must be recognized and addressed in order for these systems to be just and equal for all users.

Bias may be introduced at any point in the model creation process and can take many different forms. Fundamentally, prejudice is ingrained in our culture and the environment we live in. Bias in the world cannot be directly resolved. However, measures can be taken to minimize bias within data, models, and human review processes.

Sources of Bias:

➢ **Bias in data:**

- **Historical bias** is the prejudice that already exists in the world and has crept into our statistics. This bias tends to manifest for historically marginalized or underprivileged populations and can happen even in the case of ideal feature selection and sample conditions. In the 2016 publication "Man is to Computer Programmer as Woman is to Homemaker," the authors demonstrated how word embeddings trained on Google News articles display and actually reinforce gender-based preconceptions in society, demonstrating historical bias.
- **Representation bias**: Representation bias arises from the way a population is defined and sampled to create

a dataset. For example, Amazon's facial recognition software struggled to identify faces with darker skin tones because the training data predominantly consisted of images of white faces.

- **Measurement bias** happens while selecting or gathering labels or characteristics for use in prediction models. Easily accessible data is frequently a noisy stand-in for the true characteristics or labels of interest. Additionally, different groups frequently have different measuring procedures and data quality.

➤ **Algorithmic Bias:** Algorithmic bias is due to the errors of the implementation of an algorithm and its structure. In this form of bias, the entire system is programs it with some qualities, which lead to unfair decisions. It may result from inadequate input data or algorithm design and is recurrent as it forms part of systematic issues.

➤ **Confirmation Bias:** Confirmation bias is when a system draws findings based on implicit biases that users or system programmers may have. The system cannot recognize new patterns in the data because of this kind of bias, which restricts it to the historical trends in the data. This leads the system to not be capable of deciding based on actual merit; rather the system leads to solidifying the biases and assessments of an instance rather than evaluating whether the biases are correct or not.

Definition of Fairness:

AI and fair AI means focusing on algorithmic bias, including race, or ethnicity when it comes to decision-making by machines. As explained earlier, the meaning of fairness varies with individuals and, therefore, it is difficult to set standard. But then fairness shall include equality, justice and moral correctness, which make fairness a core component of ethical AI integration.

Types of Fairness in AI

- **Group Fairness:** Group fairness guarantees that different groups within a society are not discriminated against within the embedded Artificial Intelligence system. The primary objective of group fairness is to ensure that no group is disproportionately harmed or favoured by the AI system.
- **Individual Fairness:** Individual Fairness makes sure that comparable people are treated similarly by AI systems; the group shouldn't treat them that way. In contrast to collective fairness, this focusses primarily on the characteristics of the person. It looks at the treatment of the individuals where transparency is a major consideration.
- **Procedural Fairness:** Procedural fairness ensures that the decision-making process is transparent and equitable. This is made feasible by AI systems that make decisions in a transparent, fair, and responsible manner. Along with responsibility, this also entails algorithmic transparency and audits.

Mitigation Techniques:

Fairness and bias mitigation in AI systems are crucial as AI becomes more prevalent in essential domains like healthcare, finance, and law. As previously postulated, fairly and biasing mitigating methods seek to design solutions that can make a fair decision or can avoid systematic bias. The following is a summary of the important methods, challenges, and techniques for tackling these issues:

- **Mitigating Fairness in AI**

  Fairness in AI ensures equitable treatment for all individuals or groups, focusing on eliminating discrimination. Key approaches include:

Non – discrimination is the central concept in fairness of applications of AI in that equality is accorded to all individuals or groups of people. Key approaches include:

1. **Group Fairness:** Group fairness guarantees that different h human groups such as gender, race among others, are handled in a relative way by the AI systems.

   - **Techniques:**
     - **Re-sampling**: Filter off imbalance datasets with the help of oversampling or under sampling to give the datasets equal represents.
     - **Pre-processing:** Pre-processing data such that the use of biased representations do not distort training. (e.g., feature transformation).
     - **Post-processing:** Modify outputs so that measures fairness metrics such as equal, chance and demographic similarity are achieved.
     - **Example:** Risk **reduction** methods are meant to minimize difference between groups.

2. **Individual Fairness:** Besides, the individual fairness guarantees that similar individuals receive the same decisions, irrespective of their group membership.

   - **Techniques:**
     - **Counterfactual Fairness:** Make sure that the same decision regarding an individual is reached irrespective of sensitive attributes such as race or gender.
     - **Causal Fairness:** The use of causal inference will enable the detection as well as the management of biases at the individual level.

3. **Transparency and Accountability:** Transparency requires AI decision-making to be explained, while the accountability makes developers are the consequences of the AI system outcomes.

   - **Techniques:**

     - Using of Explainable AI (XAI) frameworks to explain the made decision.
     - regularly audits and impact assessments in order to assess violations of fairness.

- **Mitigating Bias in AI**

In this case, bias in AI is when the model reflects a form of bias by amplify pre-existing data and systemic forms of bias in the systems. Strategies to mitigate bias include:

1. **Pre-Processing Data:** Avoid bias at the data preparation level to ensure it is in line with population in great detail.

   - **Techniques:**

     - **Data Augmentation:** create artificial data for unresected group.
     - **Adversarial Debiasing:** Train patterns not to have certain biases.
     - **Diverse Data Collection:** Ensure varied source for as many different types of experiences as possible to get a wide experience of the human experience.

2. **Model Selection:** Choose some of the algorithms or models that bring about fairness and prevent discrimination within their results.

   - **Techniques:**

     - **Regularization:** penalized models for biased estimates.

- **Ensemble Methods:** Combine models to average out biases.
- **Fairness-Aware Training:** Use fairness constraints during training.

3. **Post-Processing Decisions:** Modify a model's outputs during training stage to achieve the fairness information.

- **Techniques**:

  - **Equalized odds**: Ascertain that the distribution of false positives and negatives among groups is equal.
  - **Calibrated adjustments**: They use filters to reduce the biases of the machine.

- **Challenges and Future Directions**

  - **Trade-offs:** The problem remains as how to balance fairness, with other objectives, such as accuracy.
  - **Bias Definitions:** Multi-modal fairness measures are yet to reach some consensus, and their absence contributes to the difficulties.
  - **Ethical Dilemmas**: Deciding which groups to prioritize raises moral questions.
  - **Generative AI:** Requires holistic strategies involving diverse data, adversarial testing, and continuous auditing.

Comparison of Bias and Fairness in AI

To put it succinctly, the key way to accomplish fairness is to strive to eliminate as many biases as possible from the systems to the individuals who developed the algorithm. For example, you may reduce the likelihood of unjust outcomes if you identify bias in your training datasets. Here, we can observe how each of these ideas is connected but separate in AI:

Bias	Fairness
Refers to a persistent and methodical departure of a product from its actual value or expectations.	The lack of bias and partiality in the choices made by an AI system.
might be inadvertent	Inherently purposeful and intended.
results from a number of things, such as skewed data or design.	It takes deliberate work to make sure the algorithm doesn't judge.
frequently found by examining results or trends.	ensured by proactive AI system design, monitoring, and auditing.

Understanding both concepts is crucial to ensuring that AI systems operate ethically and equitably. By addressing these issues, it is possible to eliminate negative ramifications with which AI can wreak havoc and instill people's confidence in the new models, to ensure proper utilization of the potential that they can unlock for all of the consumers. Furthermore, action that has to be taken to discuss further methods of how to achieve fairness with the help of AI has to be accepted, along with further questions of AI, including the problems of security, accuracy, and reliability of the AI system.

3.3 Ethical AI: Developing Safe and Responsible Agents

The establishment of generative AI agents is an area of great importance regarding the subject since it can help build trust and avoid harm. Ethical AI is responsible for ensuring that intelligent systems are morally sound and align with human ideals. It is also safe and just.

Figure 3.1: AI Ethics.

Source: – (geeksofgeeks 2024)

What is Ethical AI?

The idea of community discussion around ethical AI is to harm nothing and to constantly reduce the harms from AI on people and reduce the likelihood of the abuse, over-saturation, and consumer risk. Two distinct but connected ideas are sometimes referred to as "Ethical AI": worries about the morality of developing and using artificial intelligence technology, as well as worries about the morality of artificially intelligent systems' decision-making.

- **Ethical concerns surrounding the development and use of AI technology** – are mostly focused on the sociotechnical hazards that come with AI technology. Potential problems with the second kind of instances include how AI technology affects democracy, privacy, and jobs. Additionally, there is worry that when AI gets complicated, it may get out of control and become hazardous.
- **Ethical decision-making by AI systems** – These are the moral agent choices that these AI systems can make once they are put into the real world. For instance, determining who receives a loan or whether someone is freed on bail may be

done by an AI system. Since these choices may significantly affect people's lives, they must be made morally.

Elements of Ethical AI construction are:

- **Protection of individual rights:** Protective measures to preserve individual rights, such the right to privacy and an equal framework for all users, are incorporated into ethical AI models.

- **Non-discrimination in solution construction:** A number of instances of prejudice in AI-based tools and solutions have been documented. It is imperative that organizations and specialists take the initiative to lead the shift and support more compassionate responses to these dangers and problems. As the area becomes increasingly regulated in 2022, it will be crucial to develop a continuous auditing system and an intent-based solution that prioritizes causal and purposeful over the more popular correlation-based algorithm refining techniques.

- **Awareness, responsiveness, and an ongoing commitment to change:** It's not enough to just create the answer. The implementation of sufficient "space" for continuous awareness, risk management, accountability, and dedication to retraining the solution as frequently as feasible is necessary for data science teams to eliminate the possibility of bias, manipulation, and other ethical quandaries. Although this requirement might appear to be well recognized, it should not be underestimated. It is central to the idea of conscientious AI development.

Key Principles of Ethical AI:

Researchers in AI have found a number of ideas that can help direct the creation of moral AI. Although these guidelines are not yet legally binding, they can be very helpful to AI engineers as they investigate this emerging area.

Figure 3.2: Fundamental Ethical Principles in AI.

Source: – (Anglen 2024)

- **Transparency and explainability:** Stakeholders may observe the decision-making process thanks to AI transparency. This entails investigating the justifications for the decisions AI made, the information it gathered, and the algorithms it utilized. Clear outcomes foster trust, particularly when AI is seen to be able to make important judgements that might impact someone's health, finances, or even lead to an arrest. The rationale behind the AI models' decisions should be evident, as should the models themselves. Anyone utilizing or interacting with an AI system need to be able to understand the rationale behind a particular choice.
- **Accountability and Responsibility:** An important part of protecting the data is the responsibility of the organizations that handle such (personal) data. This also includes the concerns of data governance standards, which guarantee that data is handled legally and dishonestly. Those are First,

openness is crucial for the company to openly demonstrate how the AI system will operate. The organization will be solely responsible for the results of the AI system. Thus, accountability is also crucial. It is important to establish clear lines of accountability so that issues may be promptly identified and fixed.

- **Fairness and non-discrimination:** This essay makes the case that values crucial to treating everyone equally, such as nondiscrimination, should guide artificial intelligence. This encompasses both conscious and unconscious bias present in the data sets used by the AI model to make choices.
- **Privacy and data protection:** In order to create AI solutions, businesses must be mindful of user privacy and data. The capacity to honor a user's right to privacy is just as important as preventing unauthorized individuals from accessing data.
- **Safety and security:** AI systems should likely be protected from both deliberate and inadvertent abuse since everyone can turn malevolent at some time. The goal of ethical development is to design safeguards that will prevent outsiders from compromising the systems that have been developed. Every organization should incorporate security measures into the design, implementation, and deployment of AI systems in order to mitigate the risks of harm to a certain degree.

Practical Implementation Strategies of AI Ethics:

This implies the imbedding of ethical principles throughout the AI development life cycle including but not limited to the design phase, developmental phase, testing phase, implementation phase and even the ongoing monitoring and use phase.

- ➤ The application of the ethical principles takes place at the conceptualization and or design phase. AI developers need to integrate ethical thinking into the creation of an AI and more importantly encode their AI to be ethical-correct, transparent and privacy-audacious.

➢ It is also important to emphasize that it is critical to responsibly collect and control data during the developmental phase. This includes how information data sets are collected appropriately and how their storage and disposal are done rightly.

➢ AI systems should be monitored concerning performance as well as ethical standards once the system is live. It means that such problems can be detected and resolved in a timely manner, if the process of auditing is continuous.

➢ Also, making the specific function of the AI understood by users, what it can and cannot do, and what data it processes will help keep the users informed and relaxed. This can be achieved through the provision of clear documentation to the users as well as through other useful means inter alia interfaces that can enable the users to scrutinize the various choices made by the AI.

➢ Finally, the report shall show who is responsible if the system is wrong or if it results in adverse outcomes. This is a useful approach to both internal and legal responsibilities.

3.4   Transparency and Accountability in AI Systems

Fundamental of the ethical deployment of AI systems (more so of generative AI agents) is the requirement for transparency and accountability. To make sure that the understandable, the interpretable decision-making processes, and the trustworthy outputs can be ensured these are these principles. While organizations using AI must also pay attention to transparent practices and accountability mechanisms to gain user confidence and ethical compliance, the emphasis on these should be on the former.

- **Explainable AI (XAI)**

  In order to achieve transparency however, explainable AI is necessary to give insight as to how AI systems come to their decisions. Feature attribution, visualizations are just some of the techniques that can be used to explain the rationale behind an AI model's outputs to stakeholders. It also creates trust and helps find potential biases or errors in the system. XAI can be used for example in healthcare applications, by providing to clinicians a clear justification of diagnostic recommendations, thus increasing adoption and reliability.

- **Auditing and Documentation**

  Maintaining accountability in an AI system is not possible without regular audits. Audits in these cases include checking the data of the AI, its algorithms and outputs to meet ethical standards and regulatory requirements. Audits are supplemented by comprehensive documentation of design, training and deployment processes. This guarantees that everything about the development of AI is trackable, which is important to diagnose and correct the issues.

- **Ownership and Incident Reporting**

  It is important to establish clear ownership of AI systems when they are held accountable for, well, everything from the good to the bad. For this reason, organizations need to assign roles and responsibilities for monitoring, maintaining, and updating AI systems in their organization. Incident reporting mechanisms go further to hold users accountable by allowing them to report malfunctions or unethical outcomes. This involves such systems which secure prompt investigation, corrective actions, and effective control of risks and integrity of ethics.

- **Legal and Ethical Governance**

  Legal and ethical frameworks should also match transparency and accountability. Data protection laws like GDPR, along with adhering to industry standards, are the means to comply with AI system responsible operation. To avoid the hazards, organizations should perform regular impact assessments to identify possible sources of risks and social impacts and tackle the problems reactively.

- **Integration with Ethical AI Practices**

  Transparency and accountability in themselves are not aspirations but are caught up with important ethical AI practices as well. By integrating these principles into every step of AI's development from data collection to deployment, organizations will have a way to lower risk, protect fairness, and raise trust in end users. Robust governance frameworks combined with these practices facilitate responsible innovation in generative AI.

Multiple Choice Question (MCQs)

1. Which of the following is a key strategy to prevent the misuse of AI?

    A. Increasing access to all AI models
    B. Implementing robust security protocols
    C. Reducing transparency in AI development
    D. Ignoring ethical considerations

2. Bias in AI systems often occurs due to:

    A. Limited computational resources
    B. Poorly designed user interfaces
    C. Training data that reflects societal prejudices
    D. Excessive testing of AI systems

3. Ethical AI aims to:

    A. Prioritize profit over safety
    B. Reduce human involvement in decision-making
    C. Develop systems that operate responsibly and minimize harm
    D. Limit AI applications to only entertainment

4. Transparency in AI systems involves:

    A. Concealing decision-making processes
    B. Making AI systems completely open-source
    C. Clearly explaining how decisions are made by the AI
    D. Avoiding accountability for AI decisions

5. What is a common approach to ensure fairness in AI systems?

    A. Ignoring outliers in datasets
    B. Regular audits of algorithms for biased outcomes
    C. Removing all human oversight
    D. Avoiding diverse datasets

6. Which of the following is NOT an example of preventing AI misuse?

    A. Restricting access to sensitive AI tools
    B. Providing detailed training for ethical AI use
    C. Openly sharing harmful AI techniques
    D. Monitoring AI deployments

7. What does "accountability in AI systems" mean?

    A. Holding developers and organizations responsible for the outcomes of AI
    B. Allowing AI systems to operate without human oversight
    C. Making AI decisions anonymous
    D. Avoiding regulation of AI development

8. Bias in AI systems can be mitigated by:

    A. Using diverse and representative training datasets
    B. Reducing the complexity of algorithms
    C. Eliminating user feedback mechanisms
    D. Increasing the opacity of AI systems

9. Which of the following is a principle of ethical AI development?

    A. Maximizing the speed of deployment at all costs
    B. Prioritizing societal benefit over individual rights
    C. Minimizing risks and ensuring human safety
    D. Avoiding stakeholder input during development

10. What role does transparency play in ethical AI systems?

    A. It helps users understand how AI makes decisions
    B. It reduces the accountability of developers
    C. It eliminates the need for ethical guidelines
    D. It ensures that AI systems are harder to regulate

Answer

1	2	3	4	5	6	7	8	9	10
B	C	C	C	B	C	A	A	C	A

How Generative AI Agents Work

The generative AI agents have also enhanced the human-computer interaction in such ways that a number of these agents are able to perceive the environment, generate outputs as well as interact with the user. This chapter focuses on what these components essentially do and on the difficulties of improving the efficacy of their actions.

Generative AI agents are designed to create content, whether it be text, images, music, or other forms of media. They work by identifying trends in the data that is already available and then creating new material that imitates those similarities. The process involves several key components and technologies that enable these agents to function effectively (Sengar et al. 2024).

- Generative AI agents utilize algorithms to analyse large datasets.
- They learn from the data to understand context, style, and structure.
- The output is generated based on learned patterns, often with a degree of randomness to enhance creativity.

Agents are capable of supporting high-complexity use cases in a variety of business processes and sectors, especially for workflows that need specialized forms of qualitative and quantitative analysis or include time-consuming tasks. Agents do this by recursively decomposing intricate processes and carrying out subtasks across specialized instructions and data sources in order to arrive at the intended outcome. Typically, the procedure consists of these four phases (Figure 4.1):

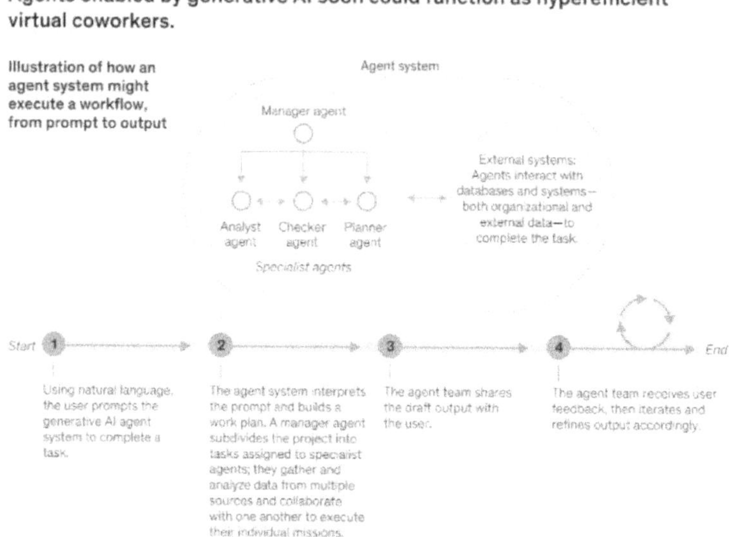

Figure 4.1: Illustrates the Step-by-Step Process Through Which Generative AI Agents Work.

Source: – (Yee et al. 2024)

Figure 4.1 Illustrates the step-by-step process through which discuss in below Generative AI agents work:

1. **User Prompt:** The process begins with the user providing a task prompt in natural language. This could be a request for generating a report, analysing data, or executing a specific action.

2. **Task Interpretation and Execution:** The Generative AI agent system interprets the user's prompt and constructs a comprehensive work plan. A manager agent supervises the whole process and then further distributes this complex task into different sub-tasks that can be easily accomplished by specific agents.

Specialist Agents:

  i. **Analyst Agent:** Compiles data from different sources and undertakes an analysis of the data.

  ii. **Checker Agent:** Responsible for maintaining quality, accuracy and validity of the information and intermediate results.

  iii. **Planner Agent:** Determines and schedules how the work processes should occur to ensure that they conform to the goal and aim.

The agents have their particular forms of tasks when interfacing with the organisational and external databases and systems. Such solicitations involve usage of resources such as databases to access relevant data, analysis data and synthesising data as may be needed.

The specialist agents are completely coordinated with each other, providing and receiving information and conclusions on the process of work, being coordinated to approach the task solution.

  3. **Draft Output Sharing:** After the specialist agents have collected and analysed the required information, the manager agent forms the preliminary result and passes it to the user for examination. This draft is the first endeavour to respond to the user's given prompt using the collected data and its interpretation.

  4. **User Feedback and Iteration:** At this stage, the user goes through the draft output and offers some comments. The agent team then works on the result to improve and optimize the work done based on the suggestions or needs of the user. This back-and-forth cycling goes on until the final result is both correct and satisfies the intended user's requirements.

In ability of the Generative AI agents to independently perform tasks, incorporate feedback and enhance subsequent performances from a series of iterations presents the agents as effective virtual employees. These agents can greatly improve the efficiency and reliability of intricate task execution across numerous applications, including data processing and writing.

4.1   Perception: Gathering and Processing Input

Perception in commercially available generative AI agents is the process of acquiring data sensed from internal and external surroundings for the sake of perceiving and 'react' to such environment. This is like our sensory that is used in decision making for an agent.

Perception in generative AI agents may be defined as the potential of an agent to apprehend signals that may be put through signal conditioning in order to enable it to influence the environment in the best manner possible. These may be text, picture or voice input signals, or other sensor inputs, depending on the nature of the agent under consideration. Their purpose is to provide semantic interpretations of the input data, that is, to identify patterns that are then used for decision making in other levels of AI systems: computer vision, NLP, and audio recognition. For example, applications such as Siri and Alexa relay on speech recognition and natural language processing to decode commands submitted by users and speak approach an appropriate response (Gupta et al. 2024).

Multi-Modal Perception: In order to preserve pertinent and significant information, multi-modal sensory fusion uses a compact state representation to produce a comprehensive view of the environment or the agent's internal state. It's true that several sensory modalities draw attention to different facets of the same experience, and when combined, they provide a more comprehensive depiction. Advanced AI agents employ complicated perception whereby information from the text, image and audio modalities is considered. This rather fuses the different aspects to facilitate not only understanding environment complexity. For instance, in soft robotics integrating vision and touch data improves interaction modality and allows robots to navigate through unpredictable environments.

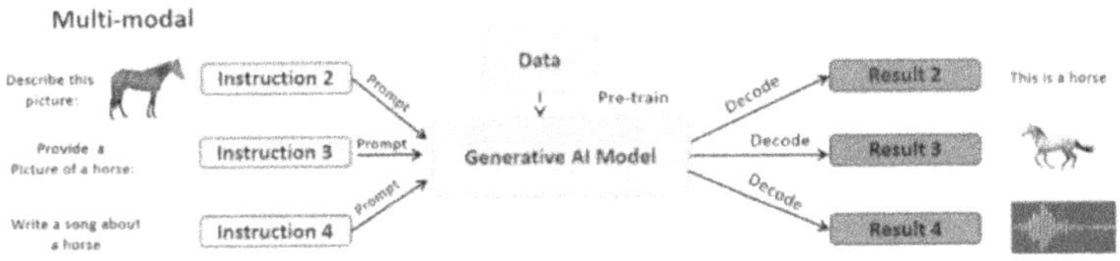

Figure 4.2: Illustrates of Highlight the Multimodal Models for Text, Image, And Audio Generation.

Source: – (Belagatti 2024)

The multimodal models, each of which represents a distinct stage, are highlighted in the diagram above. Separate models for text, picture, and audio creation are included in the model. Conversely, the multimodal model demonstrates how these disparate models are combined to create a single model that can provide material in several modalities at once. The result, you guessed it, text, visual, and audio depicting that our input image is in fact a horse!

Multimodal GenAI is based on the synergetic effect of complementing different spatial semantics, such as working with text, images, and audio. At the center of it is deep learning, an advanced aspect of artificial intelligence that tries to replicate the neurons of the human brain. These networks, achieved by a mechanism called multimodal fusion, can take information from different modalities so that the AI can understand and create content in multiple domains (Smith et al. 2025).

- **Text-to-Image Generation:** Another research direction of Multimodal GenAI is connectivity of text and image, and one of its sub-tasks is text-to-image generation. With such a structure and using technologies such as convolutional neural networks (CNNs) and recurrent neural networks (RNNs), the AI interprets textual descriptions of scenes into visual forms.

- **Image-to-Text Generation:** On the other hand, Multimodal GenAI can also generate textual descriptions for images – which is an important capability particularly in applications like image captioning and content creation. Using images as an example involving feature detection and extraction from images, the AI learns to create meaningful texts in a way that offers understanding of the inherent semantics within the visuals.

- **Audio-to-Image Generation:** Apart from text and images, Multimodal GenAI can also learn to generate images based on audio input; a feature made possible through integration of audio analysis and image generation. The AI converts audio waveforms into spectrograms and further uses generative models such as generative adversarial networks (GANs) to transpose soundscape into graphics, which expands possible ways of artistic implementation and perception for people with impaired hearing.

More recent changes in deep learning have only continued to build on the progress towards the creation of the firstly mentioned multimodal AI systems. Tools like CLIP (Contrastive Language-Image Pretraining) attend to visual data paralleling language and human understanding of the pictures. Such systems allow generative AI to be aware of the world more comprehensively and bring together gaps between different formats of data. Multimodal AI has vast potential in areas and sectors such as healthcare where, integrating patient information, images, and clinical documents results in improved diagnosis (Bayoudh 2024) deep learning algorithms have rapidly revolutionized artificial intelligence, particularly machine learning, enabling researchers and practitioners to extend previously hand-crafted feature extraction procedures. In particular, deep learning uses adaptive learning processes to learn more complex and informative patterns from datasets of varying sizes. With the increasing availability of multimodal data streams and recent advances in deep learning algorithms, multimodal deep learning is on the rise. This requires the development of complex models that can process and analyze multimodal information in a consistent manner. However, unstructured data can come in many different forms (also known as modalities.

Large Language Models (LLMs): A kind of machine learning model called a large language model (LLM) is made for tasks involving natural language processing, such language production. Self-supervised learning is used to train large-parameter language models, or LLMs, using a large volume of text. GPTs, or generative pretrained transformers, are the biggest and most powerful LLMs. Contemporary models can be adjusted for particular purposes or directed via responsive engineering. LLMs such as GPT-4 can input text that have been pre-processed through analysis of big datasets. It is used as parts of systems in the AI agents that are used in natural language processing. However, for LLMs to be able to act as agents and fully interface with dynamic environments they need to be linked to other modules like memory and planning.

Generative AI technologies, such as LLMs, have attracted a lot of interest lately. Like deep learning and electricity, LLMs are general-purpose technology that may influence a variety of applications in different sectors.

Prompts and Completion

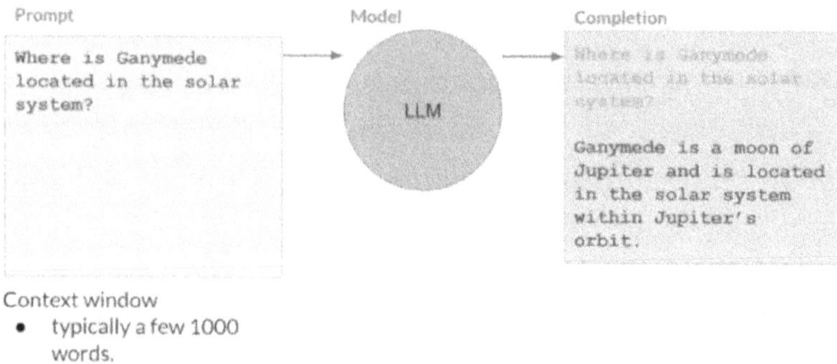

Figure 4.3: Workflow of Prompt-to-Completion Process in LLMs.

Source: – (Bhaskar 2023)

This figure 4.3 represents a basic workflow of a large language model (LLM), showing the relationship between a prompt, the model, and its completion. These models can produce language that is human-like and do intricate tasks with amazing precision because they have been trained on vast volumes of data.

A primary characteristic of LLMs is their capacity for considering context by means of a component called the "context window." I saw the use of a context window that is limited to several thousand words, helping the model remember prior information within a conversation or document. This results to appropriate and contextually sound response. For example, while asking questions such as, 'Where in the solar system is Ganymede sited?' But it's crucial to remember that the model may assist make the explanation even more accurate, thus the example that follows:

LLMs are most proficient in language processing, though using them with other modules widens their efficiency significantly. For example, combination with perception layers such as multimodal transformers allow LLMs to make a generative AI agent not only consider the textual data but also the visual and audio signals. These integrations expand their utility, hence their applicability in discordant areas such as from virtual helpers to ERPs.

4.2   Generation: Producing Outputs and Actions

Once AI agents identify input and process them, they provide the right outputs or undertakings. This generative capability is at the core of their functioning, it allows for tasks from writing articles to making decision on their own.

➢ **Generative Models:** Similar to diffusion models, generative models are used to create fresh data samples that resemble the training data. These models in AIGC provide content such as text, graphics, or music that enhances user experience through personalized work.

- **Generative Diffusion Models:** In several AIGC services, GDMs are essential approaches. DDPM and video diffusion models are used in computer vision to generate images and videos, Diffusion-LM and DiffusER are used to generate and edit text, ProDiff and DiffWav are used for audio synthesis, and GDMs are used for specialized tasks like graph generation and molecular structure creation. Deploying GDMs on network edge devices for privacy-sensitive applications is becoming more and more popular. Although GDMs may now function on edge devices thanks to advancements in model compression, the energy-intensive inference process of GDMs—that is, carrying out denoising steps—increases the total energy load on the network. Thankfully, the denoising chain design creates room for innovative distributed computing techniques.

- **Artificial Intelligence-Generated Content:** Advanced generative artificial intelligence (GAI) models are driving demand for AIGC services in fields like business and multimedia because they provide scalable and reliable text and picture output. For example, after two months, ChatGPT had over 100 million active users, demonstrating its influence on text-based communication. Multi-modal technologies have advanced significantly in the field of visual content production, as demonstrated by Stable Diffusion's ability to generate pictures from text inputs. AIGC services are widely used in human society, which suggests that Interactive AI (IAI) is the next evolutionary stage of GAI.

➢ **Context – Aware Generation:** One of the major assets of generative AI is that it is contextual. Thus, there are strategies through mechanisms such as attention layers on transformers, where models can be attentive to the most appropriate parts of the input data in order to deliver contextually relevant outputs. For example, using previous emails, the profile of the potential recipient, and the desired tone, an AI composing tool composes a well-mannered and contextually appropriate response to an email. While increasing the levels of user satisfaction this feature also aides is producing coherent and correct outputs in more intricate cases.

➢ **Autonomous Decision-Making:** AI agents have the ability to have generative capacity and make the decision on their own. By applying such LLMs and other AI models, they are able to predict actions, make decisions and accomplish them without involvement of humans. For instance, AI agents for emails, for appointments and for regulating home automation and smart facilities assists users in efficient time management.

Automated decision-making systems usually use well-known algorithms that optimize procedures according to predetermined guidelines and input data. These systems depend on a controlled, highly accurate, and consistent operating environment. The situation is made worse by the introduction of generative AI, which has a tendency to explore the possibilities and alternatives for outputs that are less likely to occur in classical AI. It implies that incorporating such technologies into decision-making processes must be done in a methodical manner.

In this sense, generative AI systems have a great chance of adding significant value to decision-making processes by developing more flexible systems. As an all-encompassing method of resource management in adaptable situations that are not supported by existing models and structures, it may also be utilized to completely utilize resources. However, when used in real-world situations, it poses unique concerns.

Personalization and Scalability: As opposed to other models generative AI is entirely unbounded, and creates material that can be personalized and personalized for a particular person. The idea is that when an AI agent is trained, it will be able to determine needs of the certain involved users and provide an output within the user necessities. For instance, an AI-based content generation writing tool can read a writer's style or simply follow a certain tone and standards of a brand. This scalability also makes AI systems to be able to work with large amount of data and thus can be advocated for in organizational uses as in content generation and business processes.

4.3   Interaction: Engaging with Users and Systems

This component defines mechanisms for interaction with users and other systems for generative AI. This phase lies in between the AI and the environment in which the created outputs make sense, are useful and easily understandable by the intended users. The current form of interaction in generative AI is complex and covers user interfaces, real time, and compatibility of AI with other systems. Therefore, interaction is an essential component of AI agents that factor in the application success of the agents. Engagement can be defined in terms of user inputs, utility of response, and as the ability to interface with other systems.

User Engagement: In these encounters, generative AI first and foremost greatly improves user pleasure and engagement. Generative AI makes meaningful and pertinent discussions possible by providing context-aware replies, which greatly connect with people and increase their level of engagement and happiness. AI agents are thus developed to be communicating with the users and make technology as easy to use. For example, the voice assistants such as Alexa, which is Amazon's assistant, is being developed based on generative AI in order to handle more extensive responsibilities, including providing users with individual Concierge that can handle multiple of their wants and needs.

Generative AI is very good at producing organic and natural interactions that closely resemble the patterns of human speech. Generative AI's sophisticated grasp of context and subtleties enables smooth and intuitive interactions, giving consumers a more human-like experience and increasing their happiness without overpowering human support representatives (Bozkurt 2023).

System Integration: The process of combining several distinct systems into a single, bigger system that works as a single unit is known as system integration. This enables companies to translate data from many sources in the technological stack and exchange it independently across various sub-systems. The introduction of generative AI (GenAI) signals a revolutionary age as system integrators traverse this changing terrain. Big language models, generative adversarial networks, and auto machine learning platforms are three essential technologies that system integrators must grasp in order to fully realize the promise of GenAI. These technologies are not as prevalent in their technical toolkit.

Thus, the abilities of the particular type of AI named LLMs — like GPT-4 — remains unparalleled in terms of text comprehension and, moreover, text creation simulating a human. With this technology in place, manufacturing systems can take instructions to a whole new level, automate documentation where necessary, and improve decision making where necessary. The system integrators should focus on building competence in NLP and make themselves knowledgeable about referencing LLMs in manufacturing systems.

Components of System Integration:

Configurations of system integration may vary depending on the integration strategy chosen but there is always collaboration between the systems in order to arrive at the specified result.

1. **Data Integration:** Integration on the other hand is the process of altering data from different sources to one format so that an organization can have a full integrated data set for analysis, reporting or data warehousing.
2. **Application Integration:** Basically, the design of systems for connectivity and cooperation of diverse applications in order to work in an integrated manner, is critical for system integration. Application integration involves establishing a unified interface that allows applications to exchange data and interact without disruptions. These applications may be from different vendors, reside on-premise or in the cloud, and have entirely different functionalities.

3. **Process Integration:** Process integration is the glue that binds the functionalities of different systems within a larger integrated system. This involves simplifying workflows, eliminating redundancies, and automating tasks to enhance productivity.

4. **Infrastructure Integration:** Infrastructure integration focuses on managing and connecting the underlying hardware and network components that support all the software applications within the system. It ensures these resources are managed effectively to support the integrated system's demands, ensuring the infrastructure can handle the processing power and data flow required by the integrated system.

Natural Language Interaction: Human interaction is most natural using language and natural language generation is one of the key strengths of generative AI. AI communicative competencies regarding textual content are enabled through technology once again in the framework of natural language processing (NLP) where agents comprehend, translate and generate functional text and language. LLMs such as GPT or Bard allow systems to perform and more conversational-based tasks, or answer questions, generate explanations or even tell stories. It is applied most commonly in virtual assistants, customer service and, to some degree, in educational platforms where the nature of interaction determines the level of satisfaction.

Multimodal interfaces: Interaction in the modern AI systems is not a mere text based but also crosses over to glass and voice, image and video interfaces. For instance, an intelligent design tool may require inputs to come in the form of images such as sketches, while the output is given as images of different prototypes depending on the changes in the design described in text. Similarly, speech-enabled devices such as smart speakers allow for interaction with the device without having to touch it because they use speech to text and text to speech. Expectations of these features make AI systems more user-friendly and flexible to meet the users' preference and demand in terms of input and output modes.

Adaptive and Contextual Responses: Two important features of Ai are user dependency and the capability to persistently keep context over several turn of conversation. flexibility in response to the user and context sustainability during multistep dialogues. AI agents use things such as context windows and attention layers to maintain a record of the prior conversation threads and have syntactic and contextual manners in their responses. For instance, in the context of a customer support flow an AI agent can remind what has been said before to improve the dialogue and avoid repeating questions. This contextual continuity makes interaction smoother and less expensive than constantly having to reinvent it.

4.4   Challenges in Improving AI Efficiency

while, generative AI has made great progress in its capacity as a tool but the optimization of that capacity remains a struggle. Demand for efficiency in AI can be defined in terms of performance, computational complexity, execution time, and the application-independent behaviour. To tackle these problems as AI systems' usage develops, scale, cost, and environmental sustainability issues need to be adapting grows. At the same time, several challenges remain to be addressed in the improvement of the efficiency and effectiveness of generative AI agents.

1. **Data Quality and Training:** The ability of AI agents to function is dependent on the availability of high-quality training data particularly that which is specific to a target task. In

contrast with other forms of artificial intelligence, AI agents need special kind of data to train and perform a particular function within particular context. Acquiring such data may often be difficult mainly because in many situations actual cases are rare or very difficult to model. For instance, training an AI agent to perform complex, nuanced tasks in dynamic environments demands detailed, context-rich data that is often not readily available. The absence of such targeted data can lead to ineffective training, where agents either fail to generalize or exhibit suboptimal performance due to gaps in their learning.

2. **Scalability and Resource Constraints:** Scalability is still an issue, its partially solved here but the biggest issue that arises is the computational power needed to process the information in real time. AI agents must run effectively on the current available hardware and energy resource which put forward a strategy that must be compensated in their performance. Generative scaling It can be difficult to scale generative AI models to manage complicated tasks and massive data volumes, particularly for smaller businesses. It can be difficult for AI models to manage complicated jobs and massive data quantities, particularly for smaller businesses.

3. **Safety and Reliability:** It is therefore important that the artificial intelligence agents are safe and reliable most of the time particularly when in use with the autonomous machinery. Some risks are related with concepts as catastrophic forgetting or with the potential of creating prejudicial results. The creation of artificial safety and ensuring that an AI system bases its actions on actual world scenarios remains an active and branching field.

4. **Human-AI Interaction:** This behaviour of human and AI interaction makes it very hard to design proper AI agents

capable to assist humans. Issues in the current scope include dependence on the self-organizing AI agents that may cause people to become isolated and the effects on the natural environment. AI agents should not overtake agents yet keeping an element of human interaction is important so as not to lose natural contact.

5. **Data Security and Privacy:** AI is made to manage vast amounts of private and sensitive data, which provide significant challenges for maintaining its security and privacy. To comply with data processing laws, businesses should have strict data protection procedures in place.

6. **Providing Coordination and Oversight:** Establishing centres of excellence (CoE) devoted to the effective adoption and distribution of emerging technologies is frequently required by organizations. When it comes to generative AI technology, these centres could be crucial.

7. **Latency and Real-Time Performance**: In the case of AI systems that operate in real-time scenarios including virtual assistants, autonomous vehicles or the fraud detection systems the latency becomes a factor. The expectations of users are high since they want quick results that may be difficult to achieve with complicated models. Latency is a critical shortage, and to minimize it while ensuring high output quality, commonplace approaches such as distillation, which trains smaller models that emulate the larger ones, and enhancing inference pathways are paramount.

In conclusion, it is possible to assert that there is great importance in developing proper operational interaction between users and the systems in the cloud-enabled business intelligence environment. By incorporating graphical interfaces, consistent system connections, and immediate feedback mechanisms, the users' experiences will be improved as well as guarantee the availability of important information. Enabling users with self-service BI and responding with interactive analytics platforms allows for a symbiotic environment of business intelligence. Lastly, focusing on user-cantered design and system improvement turns BI from a one-time exercise into a real-time, user-activation process that generates value and growth in every industry.

Multiple Choice Question (MCQs)

1. Which of the following best describes the "Perception" phase in generative AI agents?

    A. The ability to mimic human speech and gestures

    B. The process of gathering and processing input from various sources

    C. Generating textual or visual outputs based on user input

    D. Ensuring the fairness and transparency of AI systems

2. What type of data is typically processed during the perception phase?

    A. Only numerical data

    B. Text, images, audio, and other multimodal inputs

    C. Predefined outputs

    D. AI training datasets exclusively

3. During the generation phase, generative AI agents primarily focus on:

    A. Gathering user feedback

    B. Producing outputs like text, images, or actions based on input and context

    C. Ensuring fairness in decision-making

    D. Interacting with other AI systems

4. What is the main goal of the interaction phase in generative AI agents?

    A. To generate new training datasets

    B. To engage users and systems effectively, ensuring meaningful communication

    C. To detect and correct biases in AI outputs

    D. To train AI agents for real-time applications

5. Which of the following is a challenge in improving AI efficiency?

    A. Reducing computational resource requirements
    B. Increasing user reliance on AI systems
    C. Limiting AI's capability to learn new patterns
    D. Eliminating the need for real-time processing

6. In the perception phase, what role does preprocess play?

    A. Simplifying user interactions
    B. Cleaning, normalizing, and structuring input data for better processing
    C. Generating predictive outputs based on historical data
    D. Avoiding the need for manual inputs

7. What does the generation phase rely on to produce accurate outputs?

    A. User feedback alone
    B. A mix of pre-trained models, algorithms, and contextual understanding
    C. Real-time input only
    D. Manual adjustments from developers

8. What is a key challenge in the interaction phase of generative AI agents?

    A. Gathering multimodal data efficiently
    B. Maintaining user engagement while ensuring ethical responses
    C. Reducing energy consumption during training
    D. Designing neural networks with fewer layers

9. **Why is improving the efficiency of generative AI agents critical?**

   A. To limit the use of AI in complex scenarios
   B. To reduce costs, environmental impact, and latency in decision-making
   C. To make AI systems completely independent of human input
   D. To eliminate the need for training large models

10. **Which phase of generative AI agents involves feedback loops to enhance learning and interaction?**

   A. Perception
   B. Generation
   C. Interaction
   D. Efficiency optimization

Answer

1	2	3	4	5	6	7	8	9	10
B	B	B	B	A	B	B	B	B	C

PART II

Building Generative AI Agents

CHAPTER 5

Frameworks and Architectures

5.1   Agent Architectures: Core Concepts

Agent structures on the other hand, is a way of structuring intelligent agent models, which needed structures that enable autonomous decision makers, and are capable of perceiving a certain context, decide autonomously on an action to be taken and perform the action in order to meet pre-designated goals. These architectures have the central function of making architectures that enable the AI systems to function optimally in complex environments.

Rapid Innovation highlights AI agent architecture to note that it acts as a conceptual model to explain how AI systems work and how they should function. Systems need this architecture to perform tasks independently or with partial human guidance. Due to its deep expertise in AI agent design Rapid Innovation develops efficient AI solutions that work perfectly in many environments to benefit its clients. AI agents are independent intelligent computer systems that the developers create to complete distinct tasks. Business organizations employ AI agents to meet targeted goals while enhancing operational efficiency which lets staff focus on innovative work activities.

Core Components of AI Agent Architecture

Rapid Innovation views Artificial Intelligence agents as tools that read environmental data to reach target outcomes and execute needed actions. These components make up a basic AI agent setup that helps it complete its tasks according to our model below. To make AI systems and systems deliver business value organizations must fully understand these required components.

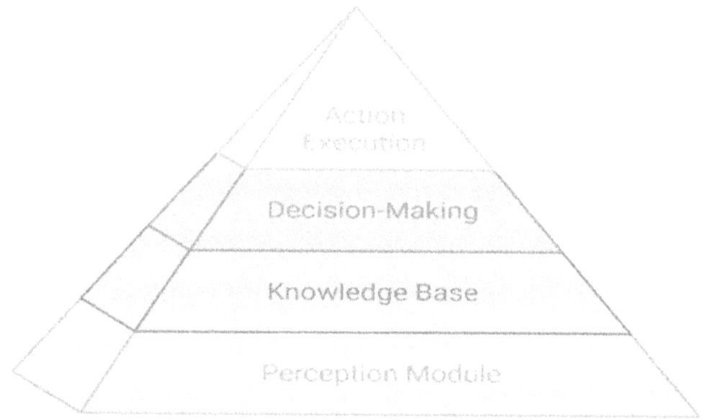

Figure 5.1: Core Components of AI Agent Architecture.

Source: – (Jesse Anglen 2024)

i. **Perception Module:**

Perception module is a module of agent's sensory system that receives and interprets the input from the external environment. This input can be text, images, audio or even video depending on the modality. This raw data then passed through the perception module, which transforms it into structured representations that the machine can understand. An example of that is the perception module, which in a text based chatbot will parse the user input using natural language processing techniques to determine intent and named entities, and sentiment or tone. Such a module might be image recognition combined with a natural language queries e.g. find an object in an image and answer related question. Downstream systems get clean, context rich data for doing further reasoning, and that is ensured by this module.

ii. **Knowledge Base:**

After understanding the input, knowledge base plays a crucial role in providing the factual or contextual grounding required for the generation of intelligent responses. Structured information included in the knowledge base may comprise with database, ontologies, or rule sets, and unstructured content (e.g., documents or conversation history) as well. Static stores include general or domain specific facts inserted in a priori and dynamic store is updated in real time for new interactions or data retrieval. Thus, in customer support agent, the knowledge base might contain product manuals, troubleshooting guides and previous support logs. To give advanced architectures the ability to support similarity search and retrieval-augmented generation (RAG), which refers to the process of having agents supplement language models with accurate and up to date information, advanced architectures often integrate vector-based memories and embedding techniques.

iii. **Decision-Making Module**

The cognitive function of the agent consists of the decision-making module. It processes the input received from the perception module, synthesizes it, based on the knowledge base, to determine an appropriate course of action. In simpler agents this might be completing predefinite rules or workflow. For more advanced systems, in particular systems with large language models or reinfornstring learning, decision making involves real time reasoning, planning and adaptive learning. The module may or may not choose to respond on its own, or may seek to invoke external API's or delegate subtasks to another agent. In addition, it measures how successful its actions were by predefined goals that can range from user satisfaction to completion of a task and then uses that feedback to refine future responses.

iv. **Action Execution Module**

The action execution module is a critical component of intelligent systems, responsible for carrying out decisions made by the system based on contextual awareness systems. This module translates high-level plans into specific actions. Like as Actuator Integration, Action Planning and Coordination, and Real-Time Responsiveness.

The action execution module converts decisions into real world actions to be followed in the environment by attaching actuators. These actuators, ranging from mechanical motors to hydraulic, are effectors that perform execute commands with high accuracy. Challenges include the ensure compatibility, reliable communication and calibration. Effective action planning and coordination align Individual and multi-agent activities synchronize within action plans. The hierarchy and algorithms help to create a plan. Increased I&O real-time responsiveness also boosts system performance by allowing timely responses to changes in the external environment or specific users' actions; it is crucial for self-driving cars and other similar applications.

8. Structure of an AI Agent

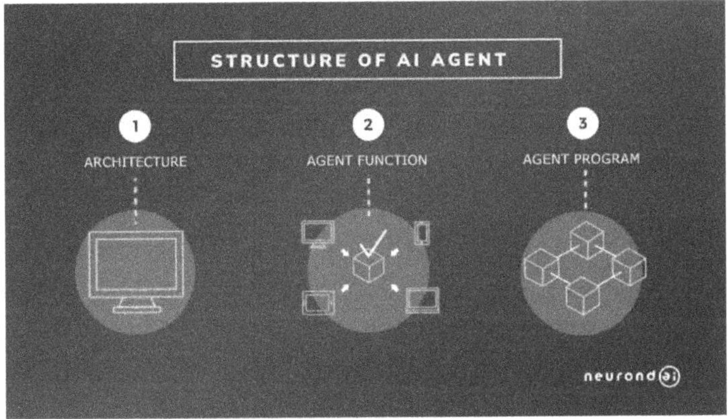

Figure 5.2: Structure of AI Agent.

Source: – (Trinh Nguyen 2024)

AI employs a variety of intelligent agent kinds, including:

a. **Simple reflex agents:** Artificial intelligence agents can be as basic as simple reflex agents. They react immediately to their current viewpoint and adhere to predetermined rules without taking past experiences or potential consequences into account.

   Key features of simple reflex agents include: Reacting to stimuli instantly, without internal memory or state, and following predetermined condition-action rules

b. **Model-based reflex agents:** Simple reflex agents are superior to model-based reflex agents. Because they have a mental picture of the world, they can take into account details of their surroundings that aren't always easy to see.

   Characteristics of model-based reflex agents: An internal representation of the environment, the ability to make decisions based on past experiences and handle situations that are not quite obvious.

c. **Goal-based agents:** An internal depiction of the surroundings, the capacity to manage issues that are not immediately apparent and basing judgements on prior experiences.

   Key aspects of goal-based agents: Clearly defined goals or objectives, the ability to plan activities, and the evaluation of potential outcomes.

d. **Utility-based agents:** To enhance the decision-making process, utility-based agents assign values to different outcomes. They aim to optimize overall "utility" or enjoyment by accounting for a range of factors and potential trade-offs.

   Features of utility-based agent: The ability to evaluate outcomes and make the best decisions under difficult circumstances while balancing several objectives.

e. **Learning agents:** Learning agents are the most advanced kind; they may acquire experience and progressively get better at what they do. These agents adjust to novel circumstances and refine their actions in response to environmental input.

   Key characteristics: The capacity to gain knowledge from experiences, consistently enhance performance, and adjust to evolving surroundings

f. **Multi-agent systems:** In multi-agent systems, several agents collaborate to solve issues or accomplish group objectives.

   Key characteristics: Communication protocols for coordination and Distributed decision-making.

g. **Hierarchical agents:** An ordered collection of intelligent agents grouped in tiers is known as a hierarchical agent. Complex jobs are broken down into simpler ones by the higher-level agents, who then allocate them to lower-level agents. Every agent operates on their own and updates their supervisory agent on their progress. The higher-level agent gathers the data and manages lower-level agents to make sure they all work together to accomplish objectives. (Alexander De Ridder 2024)

Nonetheless, AI agents often adhere to this simple structural formula: Agent Program + Architecture = Agent

The three main words used in the structure of an artificial intelligence agent are listed below:

- **Architecture:** The agent's architecture describes the system hardware it uses to perform its tasks. Robotic cars, cameras, and laptops are examples of devices that have sensors and actuators.
- **Agent function:** To connect an action with the percept sequence—a record of everything an agent has seen up to this point—an agent function is used.
- **Agent program:** An agent program operates on its physical architecture and performs the agent function.

Many artificial intelligence agents make use of a framework known as the PEAS model. PEAS stand for Performance Measure, Environment, Actuators, and Sensors. In this instance, the objective of an agent's behaviour success serves as the performance metric.

Consider autonomous vehicles as an example. This is how PEAS representation might seem in this situation:

- **Performance:** Legal drive, time, safety, and comfort
- **Environment:** Roads, other vehicles, road signs, pedestrians

- **Actuators:** Brake, horn, signal, accelerator, and steering
- **Sensors:** Camera, GPS, speedometer, odometer, accelerometer, and sonar.

5.2 Retrieval-Augmented Generation (RAG)

The AI framework known as RAG (Retrieval-Augmented Generation) blends the advantages of generative large language models (LLMs) with the strengths of conventional information retrieval systems, including search and databases. Grounded generation is more precise, current, and pertinent to your particular requirements when your data, global knowledge, and LLM language abilities are combined.

An approach known as retrieval-augmented generation (RAG) is used to improve the output of a large language model. RAG queries a trustworthy knowledge base separate from the training data sources before generating a response. Big data and billions of parameters are used to train LLMs, which then produce unique output for tasks like question answering, language translation, and sentence completion. Without requiring the model to be retrained, RAG expands the already potent capabilities of LLMs to certain domains or an organization's own knowledge base. It is an economical method of enhancing LLM output to ensure that it is correct, pertinent, and helpful in a variety of situations (Alan Zeichick 2023).

The conceptual flow of employing RAG with LLMs is depicted in the following diagram.

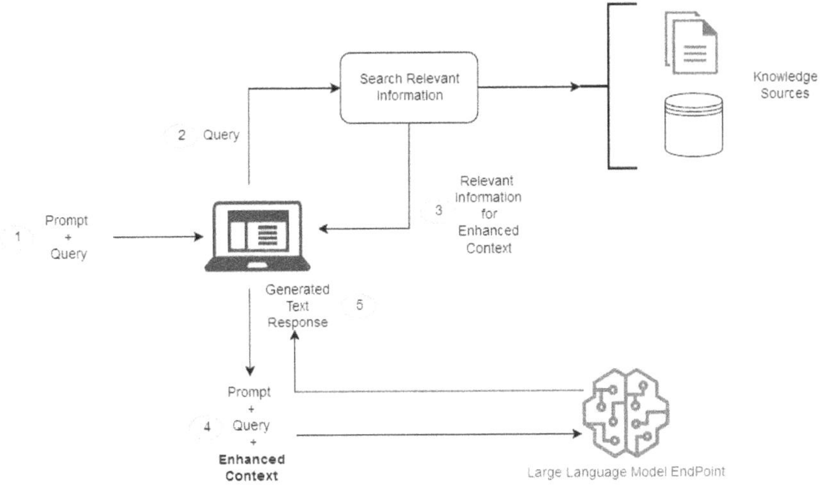

Figure 5.3: Shows the conceptual flow of using RAG with LLMs.

Source: – (RAG 2024)

Why is Retrieval-Augmented Generation important?

One of the main artificial intelligences (AI) technologies behind intelligent chatbots and other NLP applications is LLMs. The objective is to develop bots that can cross-reference reliable information sources to provide answers to user enquiries in a variety of scenarios. Regrettably, the unpredictable nature of LLM technology makes LLM replies unpredictable. The static nature of LLM training data also adds a cut-off date to the information it possesses. Well-known difficulties for LLMs include:

- Providing inaccurate information when the solution is not available.
- Providing generic or outdated information when the consumer anticipates a precise, up-to-date answer.
- Developing a response from sources that aren't considered authoritative.

- Due to terminology confusion, which occurs when many training sources utilize the same word to discuss distinct topics, incorrect replies are produced.

The Large Language Model can be compared to an overly eager new hire who will constantly respond to questions with complete assurance, yet refusing to keep up with current affairs. Unfortunately, you don't want your chatbots to adopt such a mindset because it might undermine consumer trust!

What are the benefits of Retrieval-Augmented Generation?

- **Cost-effective implementation:** The process of developing a chatbot usually starts with a basic model. API-accessible LLMs trained on a wide range of generalized and unlabelled data are called foundation models (FMs). For domain-specific or organization-specific information, retraining FMs comes at a significant computational and financial cost. RAG is a more economical way to add new information to the LLM. It expands the scope of generative artificial intelligence (generative AI) technology's usability and accessibility.
- **Current information:** It may be challenging to remain relevant, even if the initial LLM training data sources are suitable for your needs. Generic models may be updated with the most current data, statistics, or research thanks to RAG. The LLM may be immediately connected to news websites, social media feeds, and other regularly updated information sources via RAG. The LLM may then give consumers the most recent information.
- **Enhanced user trust:** Due to RAG, the LLM is able to offer accurate information with citations. Sources may be cited or referenced in the final product. Additionally, users can conduct independent source paper searches if they want

more clarification or information. Because of this, your generative AI solution could become more reputable.

- **More developer control:** RAG streamlines the testing and improvement process for chat app developers. In order to adapt to new needs or accommodate cross-functional usage, they can manage and modify the LLM's data sources. Developers may also ensure that the LLM generates pertinent responses and restrict the retrieval of sensitive data to different degrees of permission. Additionally, they may troubleshoot and resolve the issue if the LLM cites erroneous sources for certain questions. Organizations may employ generative AI technology for a greater range of applications with more confidence.

How does Retrieval-Augmented Generation work?

In the absence of RAG, the LLM uses the input from the user to provide a response based on the knowledge or information it has already been educated on. With RAG, an information retrieval component is added that initially pulls data from a new data source using user input. The LLM is provided with both the user's inquiry and pertinent data. To provide improved answers, the LLM makes use of its training data and the new information. The procedure is summarized in the parts that follow. RAGs work in a few key phases to improve the outputs of generative AI:

- **Retrieval and pre-processing:** RAGs use robust search algorithms to query databases, knowledge bases, and web pages, among other external data sources. The relevant data undergoes pre-processing after retrieval, which includes stemming, tokenization, and stop word removal.
- **Grounded generation:** The pre-processed returning data is then seamlessly included by the pre-trained LLM. This integration enhances the LLM's background and gives it a

deeper understanding of the topic. This improved context enables the LLM to generate more precise, instructive, and engaging responses.

- **Create external data:** Extern data is new information that wasn't part of the LLM's initial training set. The data might have originated from a variety of places, including databases, document repositories, and APIs. Database records, lengthy paragraphs, files, and other similar forms are all possible places to get this info. An additional AI technique called embedding language models is used to convert data into numerical representations and store them in a vector database. A knowledge library that the generative AI models can comprehend is produced by this approach.

- **Retrieve relevant information:** A search for relevance is the next stage. A vector representation of the user query is created and compared to the vector databases. Take a clever chatbot that can respond to enquiries from a company's human resources department, for instance. The annual leave policy papers and the employee's previous leave history will be retrieved by the system if an employee searches for "How much annual leave do I have?" Due of their great relevance to the employee's contribution, these particular papers will be returned. Mathematical vector computations and representations were used to determine the relevance.

- **Augment the LLM prompt:** The RAG model then adds the pertinent recovered data in context to the user input (or prompts). In order to efficiently connect with the LLM, this stage makes use of quick engineering approaches. The huge language models can produce precise responses to customer enquiries thanks to the enhanced prompt.

- **Update external data:** One could wonder what would happen if the external data became outdated. Make sure that the documents and the representation of their embedding are updated asynchronously so that the most current information is preserved for retrieval. It is possible to do this using either automated real-time methods or repeating batch processing. Data analytics frequently faces this problem; many data-science methods for change management might be applied.

5.3   Modular Approaches to AI System Design

Various modular approaches to the AI system design focus on the concept of the separation of the overall AI system into subsystems that can work autonomously of one another. Every of the modules has a particular functionality and it is more effective and efficient to develop, test as well as maintain each of these on its own. It also improves the original system's adaptability, extensiveness, and recoverability since each of the constituent segments can be modified or exchanged independently without affecting the rest of the system.(Stuart Russell 2020)

Design Patterns

1. **Pipeline Architecture:** The pipeline architecture is linear design, where data flows move through processing stages, with each stage executing a designated function. This pattern ensures clear data transformation channels and thereby makes it easy to debugging.

   Applications:

   - **Natural Language Processing (NLP):** Automated text processing techniques, part of speech tagging, opinion mining.
   - **Image Recognition:** Cleaning, feature engineering, categorisation.

2. **Service-oriented models:** Service-oriented architecture (SOA) arranges independent services for AI to work collaboratively as one unified system. Each service acts independently and integrates with other services defined by interfaces and the qualities of service provided and enabling interoperability and reusability.

Applications:

- **Microservices Architecture**: Deployment and portable AI services at different layers of cloud solutions.
- **E-commerce Systems**: real time recommendation engines and fraud detection solutions.

3. **Federated Learning Frameworks:** Federated learning frameworks allow performing of machine learning on decentralized datasets that are preserving data privacy. The model parameters are accumulated without sharing raw data to ensuring that compliance with the data privacy acts.

Applications:

- **Healthcare**: To train models without breaching patient privacy, the distributed patient records of patients across these hospitals can be used.
- **Finance**: Common fraud detection networks even in separate organizations (Anand Vemula 2024)covering its historical background, key applications across various industries, and the foundational principles underlying generative models. Readers will gain a solid understanding of machine learning basics, deep dive into probabilistic models, neural networks, and explore advanced techniques such as autoencoders, variational autoencoders (VAEs.

Challenges

- **Integration Issues:** Effective seamless communication of the various modules remains the critical challenge faced by modular designs. As with any set of products, differences in data formats, interfaces, or communication protocols have the potential to inefficiencies and errors.
- **Performance Trade-offs:** Another tradeoff is between the modularity of the model and its complexity that affects computations. Although modular designs improve on the issue of maintainability, the cost incurred in mastering inter-module communication may affect system performance.

5.4   AI Frameworks for Real-World Applications

Software platforms called AI agent frameworks are intended to make the development, deployment, and management of AI agents easier. These frameworks facilitate the creation of intricate AI systems by giving developers access to pre-built elements, abstractions, and tools. Instead of having to reinvent the wheel with each new project, developers can concentrate on the unique aspects of their apps with the help of these frameworks, which offer standardized solutions to common problems in AI agent development (Sahitya Arya 2024).

<u>Key Components of AI Agent:</u>

Important parts of AI agent systems usually consist:

- **Agent Architecture**: An AI agent's internal structures describe its decision-making mechanisms, memory systems, and interaction capacities, among other things.
- **Environment Interfaces:** Resources that connect agents to their actual or virtual workplaces.

- **Task Management:** Methods for defining, allocating, and monitoring an agent's performance on a task.
- **Communication Protocols:** Techniques for encouraging human-agent communication.
- **Learning Mechanisms:** Using a variety of machine learning approaches, agents may progressively improve their performance.
- **Integration Tools:** Tools for connecting agents to external data sources, APIs, and software systems.
- **Monitoring and Debugging:** Features that enable developers to keep an eye on agent conduct, performance, and issues.

Overview of AI popular frameworks

AI frameworks are collections of libraries that work together to make AI algorithm creation and use easier. AI frameworks are essential components for creating sophisticated, intelligent systems with the capacity to learn, adapt, and develop. They are part of any modern machine learning architecture designed with efficiency in mind (Idan Novogroder 2024).

Overview of AI agent frameworks typically include:

1. **TensorFlow:** Google developed the open-source TensorFlow deep learning framework. It offers a rich ecosystem for developing and deploying ML/DL models. Its low-level APIs provide flexibility and control, while high-level APIs such as Keras facilitate model creation. TensorFlow models can be used for various tasks, but they excel at dealing with unstructured data such as photos, audio, and text. Because of this, TensorFlow models excel at natural language processing (NLP), object identification, picture and audio recognition, and reinforcement learning.

2. **PyTorch:** The open-source PyTorch library was created by Facebook's AI research team. Together with TensorFlow, it

is one of the most well-known deep learning frameworks for researchers and practitioners. When creating deep learning models, PyTorch is renowned for its adaptability and user-friendliness, which facilitates quick debugging and intuitive model construction. A popular tool for deep learning applications such as natural language processing, computer vision, and reinforcement learning is PyTorch.

3. **Lang chain**: The creation of applications powered by large language models (LLMs) is made easier by Lang Chain, a framework that is both versatile and reliable. Developers may design strong AI agents with complex reasoning, task execution, and communication with external data sources and APIs with the aid of its numerous tools and abstractions.

   The major difficulties developers have while collaborating with LLMs include handling multi-step projects, incorporating external information, and preserving context throughout lengthy discussions.

4. **Lang Graph:** Lang Graph is an extension of Lang Chain that may be used to create stateful, multi-actor applications using large language models (LLMs). When developing complex, interactive AI systems that require multi-agent collaboration, planning, reflection, and reflection, it is extremely beneficial.

5. **Crew AI:** Crew AI is a framework for organizing AI bots that play role-playing games. In order to work together on difficult tasks, programmers might put together an AI "crew" with certain roles and responsibilities. This framework makes it easy to build collaborative AI systems that can handle complicated problems needing a variety of knowledge and coordinated actions.

6. **Kera's:** The preferred framework for many deep learning aficionados is Kera's. Kera's was developed to make neural

network construction simple and useful. It runs on top of robust platforms like TensorFlow. It provides a simpler, easier-to-use interface that is ideal for novices. Think of Kera's as a helpful roadmap to help you traverse the complex realm of neural networks. Eras is renowned for its user-friendly approach to model creation and training. It's incredibly flexible whether you're prototyping or running production code.

7. **Hugging Face:** Hugging Face is a company that specializes in natural language processing libraries and transformers. It serves as a venue for exhibiting interesting ML/DL-based applications and exchanging datasets and models. Hugging Face also provides frameworks and tools for model deployment and production optimization. AI uses Hugging Face's libraries and tools extensively for a range of text and image jobs. They are skilled in creating texts, analyzing sentiment, identifying named entities, responding to enquiries, and creating chatbots.

8. **OpenAI:** The company's state-of-the-art, pre-trained AI models, including ChatGPT, Sora, Dall-E, and others, are easily accessible to developers through OpenAI's API. With the help of this API, developers can swiftly include AI capabilities into their projects. Teams can utilise a particular Python module that makes interacting with the API easier to make things even simpler. The OpenAI API is useful for many tasks, such as producing text, photos, and audio (text-to-speech), having multi-turn conversations, responding to queries based on images, translating, and transcribing audio from supported languages to text files. Furthermore, the API can be used to fine-tune existing models to match the unique requirements of our projects.

Multiple Choice Question (MCQs)

1. What is the primary focus of agent architectures in AI?

    A. Designing hardware components for AI systems
    B. Structuring the internal organization and processes of AI agents
    C. Creating user-friendly interfaces for AI systems
    D. Optimizing cloud storage for AI data

2. Which of the following best describes Retrieval-Augmented Generation (RAG)?

    A. A technique for training neural networks faster
    B. A method combining retrieval mechanisms with generative AI for enhanced outputs
    C. A framework for organizing modular AI components
    D. A tool for data visualization in AI applications

3. What is a key advantage of modular approaches to AI system design?

    A. Increased computational speed
    B. Easier debugging and system upgrades
    C. Better aesthetic design of AI interfaces
    D. Reduced storage requirements

4. In RAG, the retrieval component is primarily used to:

    A. Generate new data based on input
    B. Retrieve relevant external knowledge for better responses
    C. Replace the generative component in AI systems
    D. Organize datasets for future use

5. **Which of the following is NOT a feature of modular AI system design?**

   A. Scalability
   B. Interoperability
   C. Monolithic code structure
   D. Reusability of components

6. **AI frameworks for real-world applications are designed to:**

   A. Focus solely on academic research
   B. Enhance practical deployment and scalability of AI systems
   C. Replace traditional software development frameworks
   D. Simplify graphical interface design

7. **Which concept focuses on the structured organization of tasks and behaviors in AI agents?**

   A. Retrieval-Augmented Generation
   B. Modular System Design
   C. Agent Architectures
   D. AI Frameworks

8. **What is the main benefit of combining retrieval mechanisms with generative AI in RAG?**

   A. Faster training of AI models
   B. Access to up-to-date and domain-specific information
   C. Elimination of generative tasks in AI systems
   D. Simplification of AI design processes

9. **In modular AI system design, interoperability refers to:**

   A. The ability to use components across different systems
   B. The capacity to run AI systems without internet access
   C. Enhanced speed of processing data
   D. Reduced power consumption during operation

10. **Which of the following best exemplifies a real-world application of AI frameworks?**

    A. Academic research in machine learning
    B. Developing AI for autonomous vehicles
    C. Writing theoretical papers on AI algorithms
    D. Teaching basic programming to students

Answer

1	2	3	4	5	6	7	8	9	10
B	B	B	B	C	B	C	B	A	B

Memory and Knowledge in Agents

6.1 AI Agent Memory

Memory is a crucial component of modern AI systems, allowing them to store and access past conversations for added context. Every LLM has a limited memory bandwidth, known as the context window. Every token present within this window is accessible by the LLM when processing an input query and helps refine the output.

AI agent systems similarly benefit from memory. Agents utilize information from various sources to complete a predefined task. Much of the key information is unavailable at runtime and must be accessed from the memory buffer to complete the decision-making process. AI Memory systems to understand all parts in represent in figure 6.1.

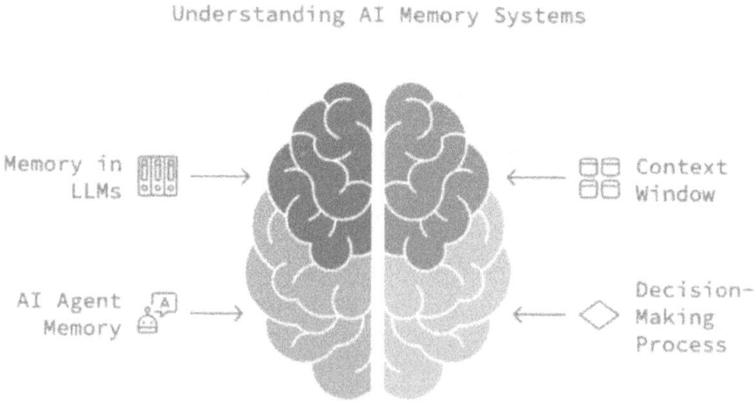

Figure 6.1: Understanding AI Memory Systems.

Source: – (Roi Lipman 2024)

What is an AI agent's memory?

A memory system allows the agent to retain crucial information from past interactions and access it later to complete the task at hand. Past information provides additional context for the present objective and improves system performance.

An individual may be capable of both working memory and working memory short-term storage. A short-term memory system only retains recent interactions and queries and is suitable in scenarios where present inputs and variables are crucial. A long-term memory system retains information over a longer time. These benefit applications by using older contexts to improve their present responses.

Why is AI agent memory important?

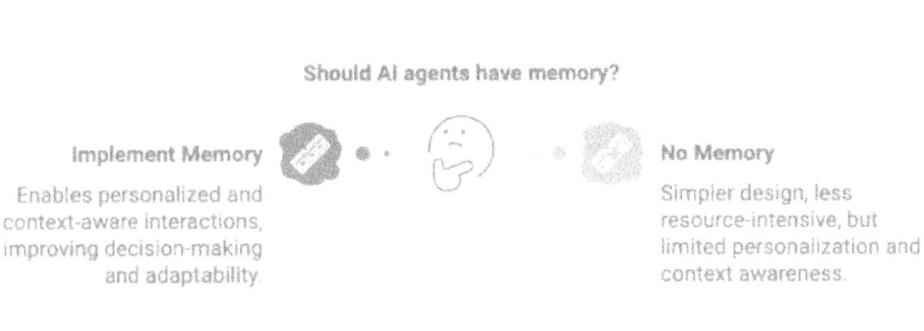

Figure 6.2: AI Agents' Memory.

Source: – (Roi Lipman 2024)

Memory plays a crucial role in enhancing an LLM's performance. It provides additional context at run-time, allowing the model to make decisions influenced by past interactions. Depending on the application, the memory module may retain a user's past preferences, sensor input data, or actions performed via actuators.

This information enhances the present decision-making process and improves the agent's response. It also allows the model to adapt to the users changing needs and display versatility and robustness in dynamic scenarios.

Type of Memory:

Memory comes in two varieties. Short-term memory comes first, followed by long-term memory. Explain in below parts.

The way that each contemporary computer handles its memory is comparable to this method. Table 6.1 below provides a clearer illustration of this parallel:

Table 6.1: Modern Computer Manages Memory: (Toward AGI 2024)

	Agents	Computers
Short-Term memory	Context Window	RAM
Long-Term memory	Vector DB, Graph DB, Relational DB, Files and SSD, External Hard Drives, Cloud Folders	Storage

6.2   Short-Term vs. Long-Term Memory in AI

AI agents can be implemented with two types of memory. These are:

1. Short-Term Memory

The short-term memory (STM) implementation acts as the model's working memory. It can retain recent pieces of information for a short time span. STM provides the agent with sufficient context to complete the current task efficiently. Once the task is finished, this memory is overwritten with new information for the next objective.

Short-term memory is ideal for use cases where the agent is to complete short tasks. The agents frequently interact with the environment, gather context, quickly complete the task, and wipe the memory. A prime use case of STMs is customer support chatbots. These bots are mostly used for short conversations. They only need to retain the present conversation and guide the user according to their query. The STM is created prior to execution by substituting data from the LTM for the pertinent variables in the prompt template. It consists of

- **Perceptual inputs**: An observation from earlier tool calls (also known as grounding)
- **Active knowledge:** produced using logic or recovered from long-term memory
- Additional crucial data from the prior decision cycle (such as the agent's current objectives).

Characteristics of STM

- Duration: Short-term memory in AI retains information for brief periods, typically ranging from seconds to minutes, depending on the task's requirements. Once the task is completed or the information is no longer needed, it is discarded or overwritten.
- Capacity: STM has a limited capacity, often compared to the "magic number seven" (plus or minus two). This indicates that it has a capacity to store around 5 to 9 bits of data simultaneously. Because of this capacity limitation, AI agents must sort through data in order of relevance, typically eliminating older or irrelevant inputs in favour of more recent ones.
- Functionality: STM serves as the immediate work area for agents of artificial intelligence. It enables users to temporarily store and process data pertinent to the current activity or engagement. Making quick decisions based on up-to-the-minute information is essential in real-time jobs, therefore this is of utmost importance.

Use Cases of STM in AI

- **Chatbots and Virtual Assistants:** Supporting conversational AI and customer service, STM records and manages the flow of a conversation. For example, STM might be utilised by a chatbot to remember the user's most recent questions and answers. Since each user input would normally be considered as a separate interaction, this enables the AI to keep context and provide consistent answers throughout the dialogue.
- **Self-Driving Cars**: As it processes sensory input in real-time, STM is an essential component of autonomous driving systems. onboard order to make split-second judgements like braking or reversing, the AI onboard the automobile use STM to follow adjacent objects, road signs, and obstructions. These judgements are based on data that has a finite lifespan; once the vehicle reaches a certain point, the data is deleted.
- **Robotic Process Automation (RPA):** Simple, repetitive activities are handled by RPA systems using STM. For example, a robot may need to fill in data fields depending on user input or respond upon instant job outcomes; both examples illustrate the necessity for these systems to process and act on real-time data.

Challenges with STM

While STM is essential for real-time operations, it has limitations:

- **Overwriting:** Given its limited capacity, STM can easily overwrite important data, especially when too much information is received at once.
- **Context Loss:** Once the data is discarded after its relevance ends, STM cannot remember past interactions, which

may lead to a lack of continuity in long-term projects or conversations.

2. Long-Term Memory

Information is stored in long-term memory (LTM) for a considerable amount of time. It can hold specific information, general knowledge, instructions, or algorithmic steps to solving a problem. There are various types of LTM and shows in figure 6.3 representation.

> **Episodic Memory:** It keeps track of all the acts' ground truth, their results (observations), and the justification (thought) for them. These occurrences can be retrieved into working memory to bolster reasoning throughout the decision-making cycle's planning phase. This can hold information regarding specific past events, such as the user's date of birth, that might have been used to solve a past problem. The same information can be used as context for a present query. It can be kept in files and relational databases. It may include:

- Pairs of input-output tools that the agent has invoked during the current run
- Historical event flows (refer to the study on Generative Agents' Memory Stream)
- Game paths from earlier installments

> **Semantic Memory:** This holds general, high-level information about the agent's environment and the knowledge obtained in past interactions. The high-level information can be reutilized to solve present problems.

It keeps track of what an agent knows about the outside world and about itself. For knowledge support, semantic memory is often initialised from an external database.

However, it may also be discovered by drawing conclusions from unprocessed data (refer to the study Reflections in Generative Agents). A few instances may include.

- **RAG:** Answering enquiries using internal documents like HR Policy or utilising data and gaming manuals as a semantic memory to impact policy.
- **RAG-based In-context Learning:** A vector database (such as Lang chain's Extending the SQL toolset) contains instructions for doing certain well-known activities that may be utilised as in-context examples for tackling more complex jobs.
- **Self-learning from user inputs**: When we reflect on our past mistakes and record the outcomes (for example, "the kitchen does not have a dishwasher"), we are using an LLM.
- **Self-learning from environment interactions:** Reflexion reflects on unsuccessful incidents using an LLM and records the findings ("the kitchen does not have a dishwasher," for instance).

➢ **Procedural Memory:** This memory is a representation of the agent's thoughts, behaviour, and decision-making processes. To bootstrap the agent, the designer must first insert the proper code into procedural memory, as opposed to episodic or semantic memory, which could be initially empty or non-existent. There are two types of it.

- **Implicit Knowledge** the LLM weights are maintained in
- **Explicit Knowledge** is expressed in the code of the agent: It is further separated into two categories:

  - Methods for carrying out actions (learning, grounding, retrieval, and reasoning)
  - Processes that carry out the actual decision-making

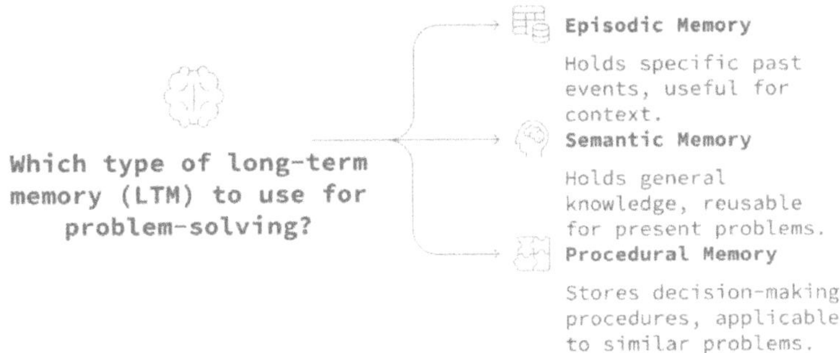

Figure 6.3: Various Type of LTM.

Source: – (Roi Lipman 2024)

Characteristics of LTM

- **Duration**: Unlike STM, LTM stores information over extended periods, from days to years. In the same way that people are able to store crucial memories for later use, AI systems may learn from their experiences and build upon what they already know.
- **Capacity**: In comparison to STM, LTM has a substantially bigger capacity. There are no real constraints on the quantity of data that AI agents may keep, including facts, rules, preferences, interactions history, and learnt behaviours. As a result, LTM is well-suited to complicated decision-making tasks that rely heavily on previous data.
- **Functionality**: AI agents are able to remember things, change their ways, and gain wisdom from their experiences thanks to LTM. It helps with more in-depth thinking by giving agents access to a wealth of information that they may use to solve complicated issues or make educated judgements.

Use Cases of LTM in AI

- **Recommendation Systems**: AI-driven recommendation systems, like those used by Amazon and Netflix, use LTM to remember user preferences and previous interactions. For instance, based on a user's viewing history, Netflix might recommend related films and TV episodes. Here, LTM is put into practice directly; the system keeps track of user data in order to provide better suggestions over time.

- **Personalized AI Assistants**: Virtual assistants that utilise LTM to recall user preferences—for example, if a user often requests the weather forecast first thing in the morning, the assistant can keep track of this and proactively deliver the information every day—to improve the user experience.
- **Adaptive Learning Systems:** Machine learning relies on long short-term memory (LTM) to train models on massive datasets. One use of AI is in fraud detection systems, where it can learn from previous mistakes and utilise that information to make better decisions going forward. This allows the system to more precisely spot fraudulent transactions.

Challenges with LTM

Although LTM is essential for retaining information over the long term, it does encounter certain difficulties:

- **Data Quality and Relevance**: Storing vast amounts of data can result in outdated or irrelevant information if not regularly updated. AI systems need robust data management strategies to ensure that LTM remains accurate and useful.
- **Complexity of Retrieval:** Accessing and retrieving the right information from LTM can be computationally expensive and complex. Efficient memory indexing and retrieval mechanisms are essential to ensure that LTM remains accessible when needed.

6.3   Using Knowledge Graphs for Knowledge Representation

Knowledge Graphs are structured representations of data that link various pieces of information to enhance understanding and facilitate data retrieval. They utilize semantic networks to represent knowledge, enabling the integration and analysis of diverse data sources, and are essential for effective graph databases.(Frank van Harmelen, Vladimir Lifschitz 2008)

In a specific area or activity, knowledge graphs (KGs) arrange data from many sources, gather information about items of interest (such as people, locations, or events), and create links between them. In AI and data science, knowledge graphs are frequently utilized to:

- Make data sources easier to access and integrate;
- Give other, more data-driven AI methods, like machine learning, context and nuance;
- provide human-readable explanations or, more broadly, enable intelligent systems for engineers and scientists to act as bridges between humans and systems.

➢ **Explaining the science**

Although comparable methods have existed since the dawn of modern artificial intelligence in fields like knowledge representation, knowledge acquisition, natural language processing, ontology engineering, and the semantic web, Google coined the term "knowledge graph" in 2012 to describe its all-purpose knowledge base. These days, KGs are widely employed in a wide range of applications, including chatbots, product recommenders, autonomous systems, and search engines. In data science, adding IDs and descriptions to data of different modalities to facilitate integration, explainable analysis, and sense-making are popular use cases. Knowledge graphs are used in AI to supplement machine learning methods in order to:

- lessen the requirement for big, labelled datasets;
- encourage explain ability and transfer learning;
- encode information related to domain, task, and application that would be expensive to learn from data alone.

A knowledge graph uses a reasoner to generate new knowledge while organising and integrating data in accordance with an ontology, also known as the knowledge graph's schema. With the help of several semi-automatic or automated data validation and integration processes, knowledge graphs can be formed from pre-existing knowledge graphs, learnt from unstructured or semi-structured data sources, or produced from scratch, for example, by domain experts.

Although there are several definitions of knowledge graphs, the most of them concur that they are:

- **Graphs**: In contrast to knowledge bases, KGs' information is arranged as a graph, with nodes—entities of interest and their kinds—and the connections and characteristics among them all being equally significant. By using links to move between sections of the graph, this facilitates exploration and makes it simple to include new datasets and formats.
- **Semantic**: An ontology, which may be represented as a schema sub-graph and identifies the types of entities in the network and their attributes, encodes the meaning of the data for programmatic application. This indicates that the graph may be used to store and arrange data as well as to infer new information and reason about it.
- **Alive**: The kinds of data and schemas that knowledge graphs can accommodate are not limited. As new information

becomes available, it is added to the graph, and they, along with their schemas, alter to reflect changes in the domain.

Certain knowledge graphs are mostly utilized by the company that developed them. The most widely used examples are Amazon's product graph and Google's knowledge graph, which are both utilized in online searches.

How do Knowledge Graphs Operate and Function?

Knowledge Graphs operate by structuring and linking data in a way that enhances understanding and facilitates knowledge discovery. They utilize a graph-based model where entities (nodes) are interconnected through relationships (edges). This style allows for a more straightforward method of comprehending complex data.

- **Data Representation**: Information Graphs facilitate the visualization of connections by representing data as a network of entities and their interactions.
- **Semantic Relationships:** They use semantic networks to define the relationships between different data points, improving context and meaning.
- **Query Optimization:** Graph databases are designed to efficiently handle complex queries that require traversing multiple relationships.
- **Scalability:** Effective graph databases can scale to manage large datasets while maintaining performance.
- **Interoperability:** Knowledge Graphs support interoperability between different data sources, enabling seamless integration of diverse datasets.

Common Uses and Applications of Knowledge Graphs

Knowledge Graphs are powerful tools that enhance data representation and connectivity across various domains. They enable organizations to leverage their data more effectively by creating semantic relationships and offering deeper insights. Below are some of the main applications in industry and technology:(Pan et al. 2024)

- **Search Engines:** Enhance search results by comprehending user intent and offering pertinent data.
- **Recommendation Systems:** Improve the user experience by making content or product recommendations based on data relationships.
- **Data Integration:** Facilitate the merging of diverse data sources to create a unified view.
- **Natural Language Processing:** Make it possible for robots to comprehend and analyze human language by using semantic information organization.
- **Fraud Detection:** Identify anomalies and connections in data that may indicate fraudulent activities.

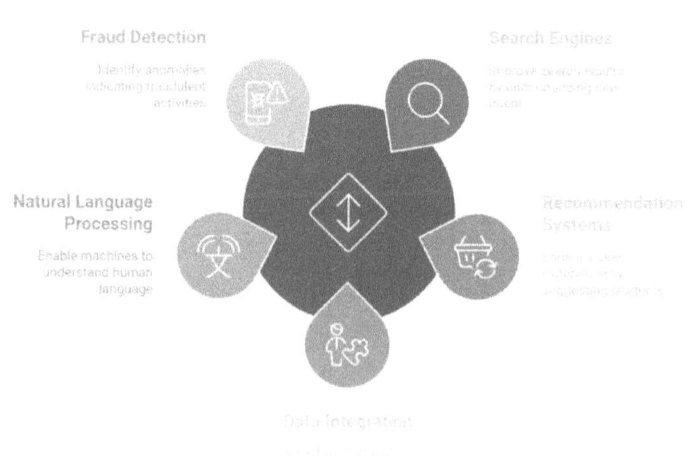

Figure 6.4: Application of Knowledge Graph.

Source: – (Lyzr Team 2024)

What are the Benefits of Using Knowledge Graphs?

Information Because graphs can describe and connect data in a way that improves comprehension and decision-making, they are extremely useful in a wide range of businesses and technology. Here are some key benefits of implementing Knowledge Graphs:

- **Improved Data Integration:** Knowledge graphs make it easier to combine many data sources and provide a thorough understanding of the information.
- **Enhanced Semantic Search:** They enable more intelligent search capabilities by understanding the context and relationships between entities.
- **Better Insights and Analytics:** Organizations may find hidden patterns and insights using linked data, which helps them make well-informed decisions.
- **Dynamic Knowledge Representation:** Knowledge Graphs can evolve over time, adapting to new information and relationships as they emerge.
- **Facilitated Collaboration:** Their ability to provide a shared understanding of data and its linkages facilitates cooperation between diverse teams.

Graph Databases for AI Agents

Graph databases are vital to AI agents because they efficiently store and query complex relationships between things through connected nodes and edges. Artificial intelligence agents may reason, infer, and make decisions by adhering to these pathways.

Agents can also perform multi-step reasoning and access updated information for real-time inference. Overall, graph databases enable AI agents to handle rich, interconnected knowledge for more intelligent and dynamic responses.

6.4 Strategies for Managing Knowledge in Evolving AI Systems

The usage of artificial intelligence (AI) is growing in popularity. Recently, companies and organizations in a variety of areas. AI has been used in healthcare, finance, and manufacturing. It has been hailed as a paradigm-shifting technology that promises to boost efficiency, productivity, and profitability. As AI technologies evolve and mature, their role in business and society will only grow (Boden 1996).

In an AI-driven world, knowledge management is crucial to leverage the power of AI, create value from data, and achieve a strategic advantage. However, organisations embracing AI must also recognise the importance of effective knowledge management. Despite AI's powerful capabilities in processing vast amounts of data and providing insights, knowledge must still be managed effectively at creation and then curated continually to retain value and increase its worth.

The Benefits of AI for Knowledge Management

AI technologies can offer several benefits for knowledge management in an organisation. AI can help to manage vast amounts of data and information quickly and efficiently, analyse and identify patterns within the data, and provide insights for decision-making. Here are some specific AI tools and techniques for knowledge management:

- **Human language and computers. Natural Language Processing (NLP)**: NLP is a branch of AI that examines interpersonal communication. NLP techniques can analyse and understand text-based Emails, reports, and posts on social media are all types of data. Natural language processing (NLP) techniques can aid in data discovery by revealing subjects, feelings, patterns, and patterns that might inform decision-making.

- **Machine Learning (ML):** Machine learning (ML) is an additional branch of artificial intelligence that deals with teaching computers to comprehend data and then use that knowledge to draw conclusions or make decisions. The capacity of ML to classify, rank, and recommend content based on user behaviour, preferences, and historical data is a boon to knowledge management. It is possible for ML algorithms to provide relevant articles and documents by studying a user's search and behaviour habits.
- **Chatbots**: Interacting with users and assisting with knowledge management tasks, chatbots are conversational agents powered by AI. In addition to finding information and answering frequently asked queries, chatbots may also tailor suggestions to each user's unique tastes and requirements.

These AI-based tactics and technology might help businesses enhance their knowledge management. To better serve its customers, IBM, for one, may use natural language processing (NLP) to sift through customer reviews and identify patterns. Consumer feedback analysis helps Procter & Gamble identify trends that might guide product creation in the future. Deep learning algorithms drive this procedure. The World Bank uses chatbots to provide customers with quick access to environmental, health, and education-related information.

Artificial intelligence (AI) has the potential to revolutionise knowledge management by supplying businesses with valuable insights and suggestions that may enhance decision-making, streamline procedures, and boost overall efficiency.

Improving AI systems makes the knowledge management problem harder to tackle. Knowledge management helps to ensure that AI agents are up-to-date and react correctly to new information

<u>Strategies for Managing Knowledge in an evolving AI-system:</u>

Here are some strategies that organisations can use to manage knowledge effectively in an Evolving AI-system: (Marc Dimmick – Churchill Felloww 2023)

- **Continuous Learning:** AI systems should be as incremental learners, the knowledge base into which the system learns should be revised as and when new data is input in the system. This approach helps to eliminate the problem of becoming outdated and instead adopt new ways of regarding models.
- **Data Quality Assurance:** It should also be emphasized that knowledge management is based on high-quality data. Appropriate methods for data moderation are cleaning, validation, and updating to maintain a qualitatively impeccable knowledge base.
- **Human-in-the-Loop Systems:** This means that implementation of human expertise in the knowledge management cycle is useful in casting an eye on data that is learned by AI systems. Thus, this partnership guarantees that the body of knowledge is in line with the intricacies and subtleties of the actual world.
- **Develop a Knowledge Management Strategy:** In order to reap the benefits of artificial intelligence (AI), businesses need develop comprehensive knowledge management strategies that back up their overarching objectives. Included in this plan should be a strategy for continuous improvement in addition to methods for acquiring, storing, and disseminating knowledge.

- **Use Value Networks:** Organisations and individuals come together in value networks to share and receive knowledge, assets, and expertise in order to create value. Organisations may enhance their knowledge management strategies and create value networks by tapping into the expertise of their associates and business partners.
- **Use Data Analytics:** Data analytics have the ability to examine information, find trends, and draw conclusions. With the use of data analytics, businesses can track performance, spot patterns, and base choices on hard evidence.

Examples of best practices for managing knowledge in an AI-driven world include:

- **Procter & Gamble:** The company facilitates collaboration and information sharing with its partners and stakeholders through its knowledge management platform, connect + Develop.
- **IBM:** In order to get insights from data, IBM's Watson knowledge management system employs machine learning and natural language processing. Both decision-making and customer service are improved by the approach.
- **GE:** General Electric uses Predix, an AI-driven system, to handle product information from design to maintenance and beyond. The system uses data analytics to find trends and make conclusions in order to optimise efficiency.

Data analytics are crucial for tracking information as AI systems evolve, allowing us to detect patterns and acquire knowledge. The ultimate objective of a comprehensive strategy that effectively utilises AI technology is to foster an environment where information exchange and collaboration are valued (Emmanuel Osamuyimen Eboigbe et al. 2023).

In order to successfully manage knowledge work in dynamic AI contexts, organisations must develop dependable systems that can accommodate both regulatory requirements and the need for flexibility. One of these is making use of AI systems' ability to learn continuously, which lets them take in new information, improve their algorithms, and adapt to their environment. This requires looking at system and user designs through feedback in order to find ways to enhance them. In a similar vein, companies should use version control and knowledge repositories to store models and data for the future. That way, when adjustments are made, decision makers will have more accurate information.

The second essential tactic is to foster collaborations between people and AI. When collaborating with AI, it's important to draw on domain knowledge to make machine learning more applicable to the organisation and its goals. One additional advantage of increasing the production of XAI systems is that it improves the cooperation between human and autonomous teams. This is due to the fact that XAI systems promote openness in decision-making, aid in the development of trust among various stakeholders, and enable efficient information exchange between human employees and AI systems. The strategies outlined above collectively maintain dynamism in knowledge management in AI systems besides being scalable and organizationally relevant.

Multiple Choice Question (MCQs)

1. What is the primary distinction between short-term and long-term memory in AI?

    A. The speed of retrieval
    B. The type of data stored
    C. The duration for which data is retained
    D. The complexity of algorithms used

2. Which of the following is a key advantage of short-term memory in AI systems?

    A. Ability to store data permanently
    B. Quick access to temporary data for real-time tasks
    C. Capability to handle complex decision-making processes
    D. Managing large-scale datasets efficiently

3. Long-term memory in AI is most useful for:

    A. Storing frequently changing information
    B. Retaining knowledge for training and future use
    C. Managing concurrent processes
    D. Performing rapid calculations

4. What is a knowledge graph in the context of AI?

    A. A graphical user interface for AI systems
    B. A structured representation of knowledge as a network of entities and relationships
    C. A visualization tool for AI memory
    D. A neural network for predictive analysis

5. **Which of the following best describes a benefit of using knowledge graphs for knowledge representation?**

   A. They eliminate the need for data storage
   B. They allow seamless integration of new data with existing knowledge
   C. They prioritize short-term memory over long-term memory
   D. They simplify all types of AI algorithms

6. **Strategies for managing knowledge in evolving AI systems focus on:**

   A. Prioritizing short-term memory over long-term memory
   B. Ensuring adaptability and scalability of the knowledge base
   C. Reducing the size of stored data
   D. Automating all memory processes

7. **What challenge is associated with long-term memory in AI systems?**

   A. Limited storage capacity
   B. Managing and updating outdated information
   C. Slow access speeds
   D. Lack of compatibility with short-term memory

8. **Which of the following is a common application of knowledge graphs?**

   A. Predicting weather patterns
   B. Powering recommendation systems
   C. Solving real-time traffic navigation
   D. Running distributed databases

9. **In the context of AI memory, "forgetting mechanisms" are important because:**

   A.   They increase the size of the knowledge base
   B.   They prevent AI systems from learning new information
   C.   They help manage irrelevant or outdated data
   D.   They slow down data processing

10. **How can evolving AI systems maintain relevance in their knowledge base?**

   A.   By constantly acquiring new data without analysis
   B.   By integrating real-time updates and pruning obsolete knowledge
   C.   By focusing only on historical data
   D.   By storing knowledge indefinitely

Answer

1	2	3	4	5	6	7	8	9	10
C	B	B	B	B	B	B	B	C	B

CHAPTER 7

Training and Fine-Tuning Generative AI Agents

7.1   Pre-Training and Fine-Tuning Basics

Generative AI agents' canonical application areas have included natural language processing, image synthesis and recommender systems. The training process of these agents typically involves two stages: pre-training and fine-tuning.

1. Pre-training

The first stage of language model learning is called pre-training. Pre-training often entails starting with the original model, randomly setting the weights, and then training the model from scratch on a few sizable corpora. Before fine-tuning a machine learning model for a downstream task, pre-training involves initializing the model on a huge, generic dataset (Jaidev and Chirayath 2012)skills and attitudes gained in training for the job. The purpose of this study is to determine whether pre-training, during-training and post-training activities significantly influence transfer of training, and also if the three predictors have a significant relationship with transfer of training. Scales developed by Saks and Belcourt (2006.

Pre-training exposes models to a large volume of unlabelled textual data from webpages, books, and articles. The objective is to record the text corpus's underlying patterns, structures, and semantic information.

More/Continuous pre-training is essentially applying transfer learning to a pre-trained model by using the weights that have already been saved from the trained model (checkpoint) and training it on a new domain (financial data, for example).

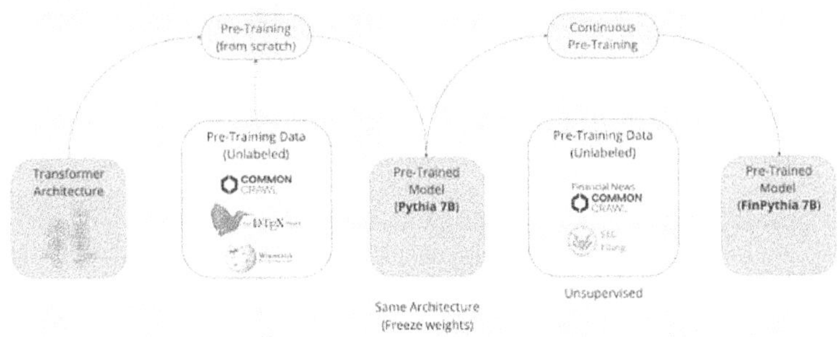

Figure 7.1: Example of further pre-train a Pythia based model.

Source: – (Ordax 2024)

The idea of transfer learning is the foundation of ongoing pre-training, as seen in the image above. A model can apply the linguistic patterns it has learnt to new datasets after undergoing basic pre-training. By using unlabelled data from a given domain, this method helps the Language Model (LLM) improve its understanding and performance in certain knowledge areas, including healthcare, law, or finance.

Key Characteristics:

- **Unsupervised Learning:** Pre-training is usually an unsupervised learning procedure in which models pick up knowledge from unlabelled text material without any labels or explicit instructions.
- **Masked Language Modelling:** In order to understand contextual links and identify language patterns, models are trained to anticipate missing or masked words inside phrases.
- **Transformer Architecture:** Pre-training often relies on transformer-based designs because they naturally understand context and link information across long text passages.

Applications:

Natural language processing demands rely on pre-trained models for tasks like sentiment evaluation text classifying and spotting named entities. These methods adjust easily to certain follow-up projects and teach comprehensive language basics.

Example and Use Cases of Pre-training:

Using substantial unlabeled text, the system learns how to interpret language data through pre-training. Training GPT-3 requires us to use millions of text documents from books, articles, and web pages. The method helps the model understand basic text patterns, internal organization, and meaning of large texts. After completing its initial training the model can then adjust for distinct use cases (H. Wang et al. 2023).

The most well-known applications of pre-training are:

- **Text Generation:** Pre-trained models are useful for applications such as chatbots, virtual assistants, and content creation because they can produce language that is coherent and contextually relevant.
- **Language Translation:** Machine translation works better when a pre-trained model gets adjusted for specific language pairings.
- **Sentiment Analysis**: By training pre-existing models with sentiment-marked data we can detect emotion in written content. Companies use these methods to collect insights from social media data and customer feedback.
- **Named Entity Recognition**: Our pre-trained models could become better at finding and pulling named entities from text-based documents.

With large unsupervised datasets Pre-training unveils the basic elements and meaning of any language type. Pre-training often includes the following steps:

- **Random Initialization of Weights:** Parameter values are first assigned at random to the model during pre-training.
- **Large-Scale Dataset:** A significant quantity of unsupervised data is used for training.
- **Learning General Features:** As a language model optimises its loss function, for example Through its cross-entropy training the model becomes better at understanding overall language structure.

Key Points of Pre-Training

- **Random Initialization**: During the pre-training phase, the model begins with a set of randomly generated parameter values.
- **Large-Scale Data:** A large unsupervised dataset is used for training purposes.
- **General Features:** The underlying structure of the language is learnt by the model. A solid grounding in grammar and semantics is necessary for moving on.

2. Fine Tuning

Using "fine-tuning" means adapting a large language model for a targeted work purpose or area through multiple learning steps. Using labeled data from the necessary topic you can start training the LLM while providing your data set. A trained LLM shows better results in its targeted area after adopting model weights for existing data. We use labeled data to teach the model unique task-specific patterns during the fine-tuning process. The model shows exceptional results compared to standard general-purpose models in its chosen area. (Anisuzzaman et al. 2025).

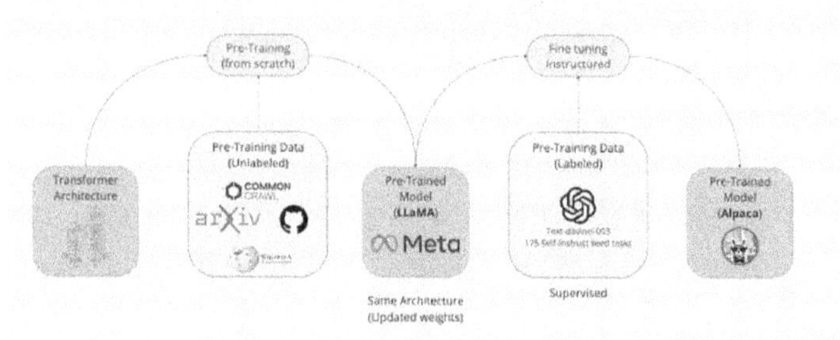

Figure 7.2: Example of fine-tuning a LLaMA-based model.

Source: – (Ordax 2024)

- **Supervised Fine-Tuning (SFT):** SFT is a way to train LLMs using labelled data and then fine-tune them. The essential labelled data consists of data sources that enter and exit connections. When supplying data into an LLM it represents input data and the desired results from the model become output data. Simple Feedback Training helps a Language Model perform better through a proven easy process.

- **Reinforcement Learning from Human Feedback (RLHF):** When powered by human feedback, LLMs receive RLHF training. Surveys interviews and user research provide multiple ways for us to gather feedback on opinions and insights from people. The RLHF training process demands greater effort yet sometimes produces results better than basic SFT systems.

Which Method Should You Use?

Several elements determine how you should refine your LLM such as how much labelled data you can access along with the time availability and resources at hand and ultimate performance targets. Labeled data fits well with SFT as an optimal solution. If you need performance upgrades beyond SFT's capabilities or lack data labels SFT can still be replaced by RLHF.

Key Characteristics:

- **Transfer Learning:** Applying learned representations from pre-training allows us to refine models for a specific target task during fine-tuning.
- **Task-Specific Data:** Specific workplace data with sentiment tags or QA pairs helps train our models.
- **Gradient-Based Optimization:** Our model parameters need gradient-based methods to adjust themselves so they work better with specific task data.
- **Applications:** Specialized natural language processing tasks including text generation machine translation question answering and sentiment analysis deliver high results when models receive proper adjustments.

Example and Use Cases of Fine-Tuning (SFT or RLHF):

Machine learning algorithms learn better when developers input labelled task data before fine-tuning pre-trained language models.

Example: It can prepare BERT for use by feeding it a database of customer reviews rated as positive or negative. Training with these texts makes the model detect the review authors' true emotional states effectively.

Use Cases:

- **Sentiment Analysis**: Sentiment analysis improves in tasks such as marketing studies when models are further trained for specific uses.
- **Text Classification:** Fine-tuned models show their value by sorting textual information into subject categories plus classifying documents and filtering out spam. These models allow text input to be automatically placed in designated categories.

- **Question Answering:** System models that adapt question-response connections through context enable effective information retrieval and customer service functions.

Fine-Tuning To enhance model performance on one task people often create specialized datasets for that task. The standard tuning process typically has these core actions at work:

1. **Loading Pre-Trained Weights**: As a first step in expediting deep learning tasks, our model incorporates bias values and weights taken from a previously trained model.
2. **Task-Specific Data**: We use a training dataset containing only activity-related data.
3. **Optimizing Task Performance**: The model needs improvements so you tweak its inner workings by optimizing loss data metrics.

Key Points of Fine-Tuning

- **Loading Pre-Trained Weights**: In order to keep the general characteristics learnt during pre-training, the model's parameters are loaded from the pre-trained model.
- **Task-Specific Data**: A dataset tailored to each task is used for training.
- **Task Optimization**: In order to optimise performance on the specific task, the model's parameters are further adjusted.

Challenges in Pre-Training and Fine-Tuning

While effective, these processes come with challenges:

- **Resource Intensity**: Each stage is highly computational in resources.
- **Data Quality:** It clearly showed that poor quality or even biased datasets can adversely affects outcomes.

- **Overfitting:** Because fine-tuning uses small training samples it can produce results that suit the training data but fail for real-life use.

7.2   Reinforcement Learning for Advanced Capabilities

As far as academic fields go, Reinforcement Learning is at the cutting edge. This innovation allows AI to mimic the way humans learn by making mistakes and trying again. By employing this approach AI learns to handle challenging games checks like Go and Dota 2. The system's response after rewarding or punishing agents updates their behavioral patterns each moment to make them better performers. Steady Performance Understanding helps AI make choices based on real-world fluctuations when they receive reinforcement learning training.

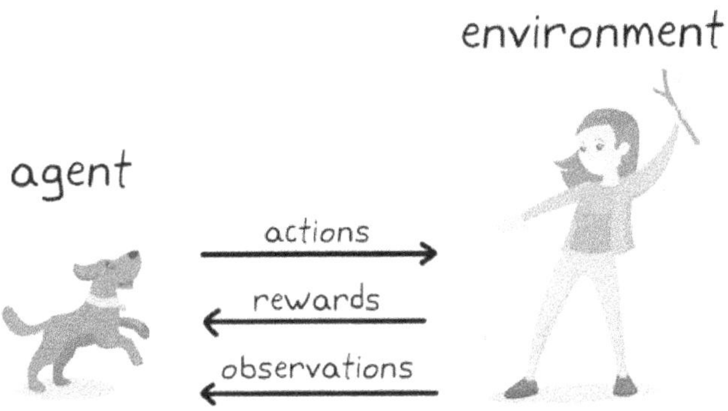

Figure 7.3: Reinforcement Learning Example.

Source: – (Prathima 2024)

Machine learning includes reinforcement learning as a subset. In this scenario, agents come up with their own reward and punishment schemes. In order to maximize benefits and minimize punishments, it entails making observations in a specific situation and selecting the best course of action. It signals behaviours that are both positive and detrimental. In essence, an agent—or several—is built with the capacity to see, comprehend, act upon, and interact with its environment.

To comprehend the concept of reinforcement learning, let's go over its formal definition.

A subfield of machine learning called reinforcement learning involves agents acting in a way that maximises their total rewards. The NVIDIA

The foundation of reinforcement learning (RL) is rewarding positive behaviour and penalising negative behaviour. The method is taught to choose the best result depending on a number of variables, and it generates several outputs rather than a single result from a single input. – The Gartner

The Framework for Reinforcement Learning

In this part, the fundamental framework of reinforcement learning will be described:

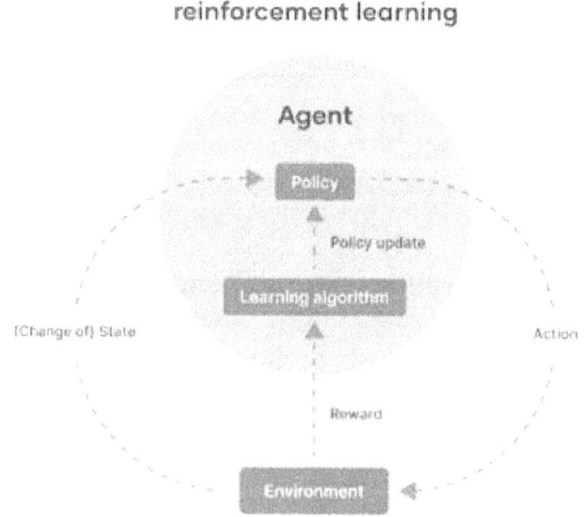

Figure 7.4: The General Framework of Reinforcement Learning.

Source: – (Bhutanadhu 2023)

- **The Acting Entity: The Agent:** An "agent" is the computer model in the field of artificial intelligence and machine learning that is given the responsibility of interacting with a certain external environment. Its main objective is to make choices and execute activities that either optimize rewards by following a set of stages or achieve a predetermined objective.
- **The World Around: The Environment:** There is an external system or setting called the "environment" in which the agent operates. Basically, this encompasses everything that can be seen but cannot be changed. A real-life example of this may be a robot navigating a labyrinth, or from an interface in a virtual game. To assess the agent's performance, the "ground truth" is the environment.
- **Navigating Transitions: State Changes:** In reinforcement learning the entity "s" represents all the adjustable situations that an agent encounters while operating in its environment. By understanding state changes, agents discover useful data about their observations, which helps shape their future actions.
- **The Decision Rulebook: Policy:** The "policy" represents what actions an agent will perform based on existing states. As a tool that converts state information into action choices, a policy details the steps required for success.
- **Refinement Over Time: Policy Updates:** The term "policy update" represents the regular process of changing what the agent does today. With this flexible reinforcement learning system agents can adjust their actions based on both old and new information they have received. Dynamic algorithms help the agent change its approach.
- **The Engine of Adaptation: Learning Algorithms:** By using learning algorithms, the agent develops the mathematical knowledge required for better policy decisions. Depending on

the situation, these algorithms may be roughly divided into two categories: The learning approach splits into model-based which uses an exact simulation of the learning environment and model-free which studies real-world interactions.

- **The Measure of Success: Rewards:** Ultimately the environment provides specific feedback based on how well an agent's actions work right now. Our evaluation system tracks how well the agent combines rewards across every time step.

Agents conduct interactive learning by constantly growing relationships with their environments. Through feedback incentives the agent tests out multiple action decisions to apply a defined policy. Our policy improves step by step with learning methods because the agent moves toward best results within allowed environmental conditions.

Real-World Applications

By joining generative AI with reinforcement learning techniques, we uncover many new options. Let's look at some real-world examples:

1. **Content Generation:** The power of AI content generation keeps getting better at matching exactly what users want to see. Visualize an RL system making a personalized news feed from GPT-3 input. After reading an article each user provides feedback. The system works by changing user feedback into numeric values through the selection of like or dislike buttons.

2. **Art and Music:** Artificial intelligence (AI) techniques allow artists to create music and artwork that reacts to human emotions by adapting its style to suit different audiences. In order to make media that resonates with people on an emotional level and changes its style depending on what viewers think, RL agents can tweak neural style transfer algorithm parameters.

3. **Autonomous Vehicles:** The AI algorithms of self-driving cars have the potential to improve both their efficiency and safety by learning from real-world driving situations. A number of variables could prompt a real-time course correction from an RL agent in an autonomous car. These include time, fuel efficiency, safety, and weather.

7.3   Transfer Learning for Industry-Specific Use Cases

Transfer learning is the process of knowledge being transferred from one field of study to another. Machine learning and AI experts use the term "model reusing" to describe the practice of applying previously trained models to new tasks.

Training convergence is accelerated since the model has learnt broad characteristics from several datasets, so it needs fewer iterations to transfer learning and adapt to the details of a new job. Shorter implementation and iteration cycles made possible by this acceleration will be of particular use to AI engineers working on time-sensitive projects.

The concept is akin to how humans learn new skills. Let's take an example: Imagine you are an accomplished guitar player and decide to learn the ukulele. Your prior experience with the guitar will accelerate your learning process. This is because many of the skills and knowledge required for playing the guitar—such as finger positions, strumming patterns, understanding the fretboard, music theory, and rhythm—are also applicable to playing the ukulele.

Why is Transfer Learning Used?

The use of transfer learning techniques in neural network development has numerous strong arguments in its favour, like as:

- **Training Efficiency**: Transfer learning shortens training time in two ways: first, it eliminates the need to develop models from scratch. Second, it allows fine-tuning with fewer datasets.

- **Model Performance:** To reduce overfitting and enable faster and more effective training with less data, transfer learning enhances model performance by using previously taught knowledge.
- **Reducing Operational Costs:** Transfer learning saves costs for training models because it eliminates the need to obtain data and buy computational resources.
- **Enhanced Adaptability and Reusability:** Using transfer learning improves how models work and achieve their benefits across different environments and projects.

How Transfer Learning Works:

Let's have a look at three ideas connected to transfer learning, Transfer learning helps people develop multiple skills while learning their nature and specific ability.

- **Multi-task learning:** Individual models learn multiple tasks by training one model for all jobs simultaneously. Following basic input processing the model splits into different layers for each work task. As the model learns related features for every task and also learns specific characteristics for task uniqueness. Current LLMs rely on this training structure. Our Introduction to LLMs Course can teach you more about this subject.
- **Feature extraction:** By using existing models we can extract useful data features from our inputs. A new model, this one tuned to a particular task, takes these characteristics into account. Feature-based transfer learning leverages neural networks' unique capabilities to extract features from data. With feature extraction, the model 'figures out', so to speak,

what part of the input is important to, for example, classify an image. This effectively means that only the initial and intermediate layers of a model that have already been trained and contain the more generalisable knowledge are used when the model is used for a new job.

- **Fine-tuning:** Fine-tuning goes beyond feature extraction and is commonly used when the two tasks are not closely related. It entails using a domain-specific dataset to further train an already-trained model. The majority of LLM models available today perform quite well overall, however they frequently fall short in certain task-oriented issues. The model is fine-tuned to perform better for particular tasks, increasing its effectiveness and adaptability in practical applications. If you want to know more about fine-tuning, check out our Introductory Guide to fine-tuning LLMs.

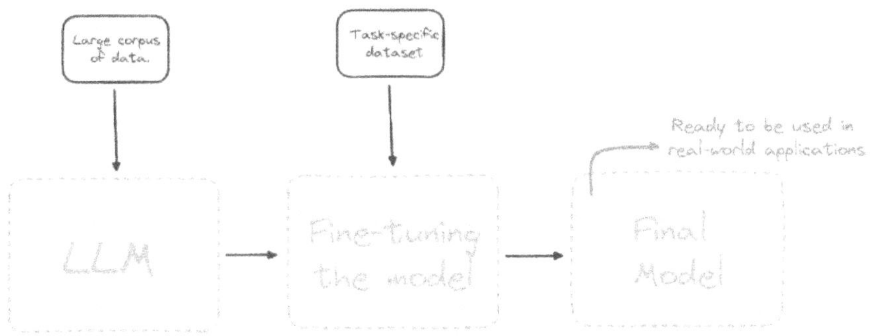

Figure 7.5: Visualizing the fine-tuning process.

Source: – (Ferrer 2024)

Applications and Use Cases of Transfer Learning

A popular method for handling many data science problems, such as those in computer vision and natural language processing, is transfer learning.

1. **Computer vision:** Computer vision is one of the fields where transfer learning has been especially fruitful. Neural networks developed in this field require vast amounts of data to address tasks like object detection and image classification. Today, there are a good number of pre-trained, public neural networks in computer vision, for example:

   - **Visual Geometry Group:** A popular starting point for many picture classification problems is the VGG model, which has already been trained.
   - **You Only Look Once:** The state-of-the-art, pre-trained object identification model YOLO is exceptional in terms of speed and broad object recognition capabilities.

2. **Natural language processing:** As a branch of AI, natural language processing (NLP) studies how computers understand and use human language. The goal is to teach computers to interpret and evaluate vast volumes of textual or audio natural language input. The applications of NLP are practically endless and include voice assistants and speech recognition.

   Transfer learning is behind some of the most popular NLP models, including:

   - **Bidirectional Encoder Representations from Transformers:** One of the earliest transformer-based LLMs was BERT, developed by Google researchers in 2018. It's a public, pre-trained model that excels in many NLP problems, including language modelling, text classification, and machine translation.
   - **Generative Pre-Trained Transformer**: At the front of the GenAI revolution, GPT is a suite of potent LLMs that were created by OpenAI. The basis models for the well-

known ChatGPT are GPT models such as GPT-3.5 and GPT-4. Additionally, OpenAI just released GPT-4o, the most potent version to yet.

3. **Self-learning Robots**: The obvious solution option, which involves having process specialists improve the simulation, is frequently impractical. Domain randomisation provides a different and far more effective method. To generate fresh training scenarios, we may now randomly alter the simulation's settings. An AI agent may be trained to withstand parameter fluctuations by using a distribution of simulations, even though the resulting simulations may differ significantly from the actual process. This means that it is now possible to integrate an AI agent into a real-world process without having to observe the exact training conditions.

4. **Wear Prediction**: The initial use of continuous learning techniques for wear prediction was regularisation on a dataset comprising turbine data. When training for the target problem, regularisation methods impede the modification of parameters vital to solving the source problem. As a result, we looked at a use case for estimating lithium-ion battery deterioration and evaluated several regularisation algorithms. There is a lack of specificity about the physical (wear) processes, and the use case encompasses a wide range of practical applications. Using regularisation techniques substantially increases mean prediction accuracy (from 54% to 70%), regardless of the significance of the individual (sub)problems' sequence and similarity.

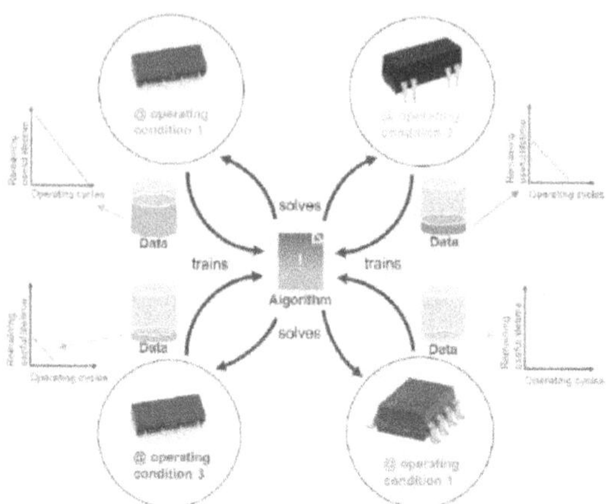

Figure 7.6: Cross Scenario Learning for Efficient Data-Based Wear Prediction on Small Datasets.

Source: – (Maschler et al. 2022)

7.4   Fine-Tuning Techniques and Optimization Challenges

Let's talk about two typical methods for fine-tuning: training with human feedback (RLHF) and supervised tuning methods:

1. **Supervised fine-tuning techniques:** This approach trains the model with a task-specific tagged dataset; in this dataset, each input data point is associated with a label or the correct answer. The model learns to adjust its parameters in order to get the most precise predictions for these labels.

   a. **Basic hyperparameter tuning:** To fine-tune the model's performance, just manually adjust its hyperparameters, which include the learning rate, batch size, and epoch count.

   b. **Transfer learning:** When there is not enough task-related data transfer learning performs better because it works well with these types of needs. Our technique begins with a pre-trained model that learned from a significant dataset of general data.

c. **Multi-task learning:** We refer to teaching a model multiple related tasks as multi-task learning. Our goal is to boost model performance by using this model to solve multiple connected tasks. Multitasking helps a model better understand data while producing higher-quality results.

2. **Reinforcement learning from human feedback (RLHF):** Researchers now combine standard language-model training methods with human feedback through reinforcement learning. The system updates its language model regularly through RLHF because it takes human feedback during every learning phase. The system generates better results which match actual circumstances.

   a. **Parameter efficient fine-tuning:** To improve the performance of pre-trained LLMs on certain downstream tasks, our strategy employs parameter-efficient fine-tuning (PEFT), which involves adjusting a small subset of model parameters. The system runs training more effectively because it only changes a portion of model parameters.

   b. **Reward modelling:** The technique produces multiple outcome alternatives and lets individuals rank these options by assessing quality. After recognizing what induces people the model adopts new ways to optimize its results.

   c. **Proximal policy optimization:** The language model's policy adjusts repeatedly using PPO to find better reward optimization. In PPO we adjust policies step-by-step to achieve better performance. Our approach puts constraints on policy changes that permit minor improvements while stopping major changes that could harm the system.

Challenges of fine-tuning Models

1. **Risk of Overfitting:** When you adjust your model too closely to fresh data it becomes less adaptable. Regularisation and cross-validation techniques help us maintain the model's effective performance with unmarked inputs.
2. **Resource Constraints:** Big models use a lot of resources when their parameters need adjusting. Cloud-based computing tools and better model layouts help reduce system demands effectively.
3. **Hyperparameter Complexity:** It remains tough to choose suitable hyperparameters but this selection stays a vital step for the model. Automatic parameter optimization tools in Bayesian methods simplify setting proper hyperparameters in machine learning models.

Common fine-tuning use cases

The application of fine-tuning enables us to customize our models for specific tasks reinforce core knowledge and broaden their scope.

- **Customizing style:** Models may be adjusted to represent a brand's intended tone in a variety of ways, from including intricate behavioural patterns and unique drawing styles to making little changes like starting every conversation with a kind greeting.
- **Specialization:** LLMs' broad language skills can be refined for certain jobs. As an example, Meta's Llama 2 models were released in three different flavours: Llama-2-chat, code Llama, and base foundation models.
- **Adding domain-specific knowledge:** Despite being pre-trained on a vast amount of data, LLMs are not infallible. In contexts like law, finance, or medicine, where using

specialised, esoteric language that may not have been well represented in pre-training is common, using extra training samples to augment the underlying model's knowledge is especially pertinent.

- **Few-shot learning:** It is frequently possible to refine models with high generalised knowledge for more specialised categorisation texts with relatively few illustrative instances.
- **Addressing edge cases:** You could want your model to respond in a particular way to specific scenarios that were probably not addressed during pre-training. One efficient method to make sure these circumstances are handled correctly is to fine-tune a model using labelled samples.
- **Incorporating proprietary data:** It's conceivable that your company has a data flow that is exclusive and very relevant to your application. To start the process the model small logic can begin incorporating this dataset into its training rather than building the knowledge base from scratch.

Businesses need to match AI model performance with efficient resource use to make their AI models work better and resolve optimization problems. Companies must balance underfitted and overfitted models to make sure their design works with different datasets. Using transfer learning models for multiple jobs helps us save training time while keeping accurate results. Models improve at their assigned tasks according to both grid search and Bayesian optimisation hyperparameter designs. You need an orderly plan to deal with problems from biased data plus machine learning expenses plus smaller progress returns. A business can improve its AI systems both now and in the future by checking model output and adding real-world data regularly.

Multiple Choice Question (MCQs)

1. **What is the primary goal of pre-training in generative AI models?**

    A. To optimize the model for specific tasks
    B. To fine-tune the model on domain-specific data
    C. To learn general patterns and knowledge from a large dataset
    D. To eliminate bias in AI systems

2. **Fine-tuning is typically performed after pre-training to:**

    A. Reduce the model size for deployment
    B. Adapt the model to specific tasks or domains
    C. Remove redundant layers in the neural network
    D. Increase the size of the training dataset

3. **Which of the following best describes reinforcement learning in AI?**

    A. Learning from labeled datasets to classify data
    B. Training the model by maximizing rewards for desired outcomes
    C. Using pre-trained models to speed up training
    D. Optimizing the model for industry-specific tasks

4. **A key advantage of transfer learning is:**

    A. Reducing computational costs by reusing pre-trained models
    B. Ensuring the AI model is free from bias
    C. Generating entirely new datasets for training
    D. Optimizing neural network architecture

5. What challenge is often encountered during fine-tuning?

- A. Lack of pre-trained models
- B. Overfitting to the specific dataset
- C. Insufficient generalization during pre-training
- D. Limited access to reward functions

6. Which method is most commonly used to adapt pre-trained models for new tasks?

- A. Transfer learning
- B. Self-supervised learning
- C. Zero-shot learning
- D. Evolutionary algorithms

7. Reinforcement learning for generative AI agents often involves:

- A. Backpropagation with supervised labels
- B. Feedback loops to maximize cumulative rewards
- C. Training on static datasets
- D. Using adversarial networks for improved accuracy

8. In transfer learning, the process of using a pre-trained model as a starting point for new tasks is called:

- A. Reward optimization
- B. Domain adaptation
- C. Knowledge distillation
- D. Model initialization

9. A major optimization challenge in fine-tuning is:

- A. Selecting the right hardware for training
- B. Balancing generalization and task-specific performance
- C. Increasing the size of the dataset
- D. Choosing a suitable unsupervised learning algorithm

10. Which technique is often used in reinforcement learning to explore new strategies?

A. Supervised learning
B. Gradient clipping
C. Exploration-exploitation trade-off
D. Batch normalization

Answer

1	2	3	4	5	6	7	8	9	10
C	B	C	B	A	A	B	D	B	C

CHAPTER 8

Enhancing Agent Capabilities

8.1   Natural Language Understanding and Generation

Natural language generation (NLG) and natural language understanding (NLU) are two different but related fields. NLU and NLG are only parts of NLP at a broad level (Raiaan et al. 2024).

What is natural language understanding?

As a subset of natural language processing, natural language understanding analyzes voice and text using syntactic and semantic analysis to ascertain a sentence's meaning. Semantics suggests the intended meaning of a statement, whereas syntax describes its grammatical form. A pertinent ontology, or data structure that outlines the connections between words and sentences, is also established by NLU. The combination of these analyses is necessary for a machine to comprehend the intended meaning of various texts, even if humans naturally accomplish this in conversation. (Khurana et al. 2023).

It facilitates the machine's comprehension of the data. To handle data appropriately, it is utilized to interpret data and determine its meaning. By comprehending the context, semantics, grammar, purpose, and sentiment of the text, it resolves the issue. Numerous guidelines, methods, and models are employed for this aim. It determines the text's purpose. Three linguistic levels exist for language comprehension. (Gill 2024).

- **Syntax:** It can comprehend phrases and sentences. It examines the text's syntax and grammar.
- **Semantic:** It verifies that the text makes sense.

- **Pragmatic:** It comprehends context to determine the purpose of the text.

In a structured and proper format, it must comprehend the unstructured text with errors. It transforms text into a format that can be read by machines. Semantic analysis, conversation agents, semantic phrasing, and more applications use it.

Unlike other NLP tasks that might involve generating or translating text, NLU is particularly interested in comprehending the context, meaning, and intent of the words and phrases that people use.

Figure 8.1: Best approach to understanding human language.

Source: – (Dimitri Didmanidze 2024)

At its core, NLU is about transforming unstructured language data into structured information that machines can work with. This involves tasks such as identifying entities in a sentence, determining the sentiment of a statement, or classifying the intent behind a user's query.

The subtleties of language are aptly demonstrated by our capacity to discriminate between homophones and homonyms. Let's consider the next two statements, for instance:

- Alice is swimming against the current.
- The current version of the report is in the folder.

The term "current" is a noun in the first sentence. We may infer that we are talking to the movement of water in the ocean since the word swimming, which comes before it, gives the reader further information. The term current is used as an adjective in the second phrase. Version, the word it defines, indicates several iterations of a report, allowing us to ascertain that we are speaking of the most recent version of a file.

How NLU Works

Natural language understanding (NLU) relies on a combination of linguistic rules, machine learning algorithms, and statistical models to interpret human language accurately. (Gumińska, Poniszewska-Maranda, and Ochelska-Mierzejewska 2023).

NLU systems transform unstructured language data into structured information that machines can use to perform tasks such as sentiment analysis, entity recognition, and intent classification.

Modern NLU relies mostly on deep learning, particularly models like transformers. For example, even multi-purpose models like ChatGPT are capable of parsing any English language sentence to extract the meaning. (Samant et al. 2022).

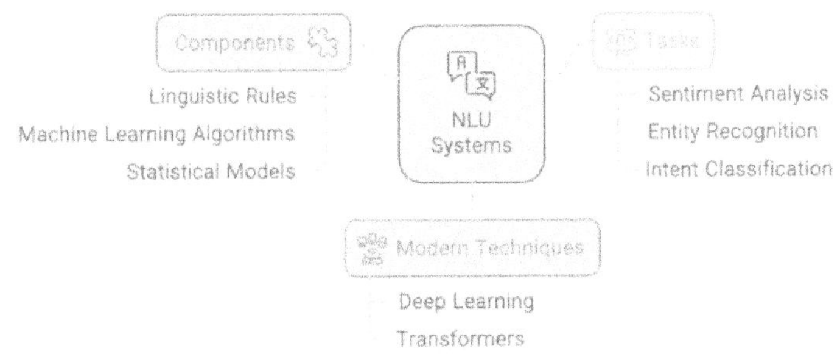

Figure 8.2: NLU Systems.

Source: – (Dimitri Didmanidze 2024)

What is natural language generation?

Natural language processing also includes natural language production. Computer reading comprehension is the emphasis of natural language understanding, but computers can write thanks to natural language generation. The process of creating a written answer in human language using NLG is dependent on input data. With text-to-speech services, this text may also be transformed into a spoken format. (Perera and Nand 2017).

Natural language generation (NLG) is a method for creating meaningful phrases. With a fast speed of hundreds of pages per second, it simplifies the structured data for human comprehension. The following is a list of some of the NLG models:Natural language generation (NLG) is a method for creating meaningful phrases. With a fast speed of hundreds of pages per second, it simplifies the structured data for human comprehension. The following is a list of some of the NLG models:

- Markov chain
- Recurrent neural network (RNN)
- Long short-term memory (LSTM)
- Transformer

NLG additionally has text summarisation features that preserve the information integrity while producing summaries from input documents. Key Point Analysis in That's Debatable is powered by an AI invention called extractive summarisation.

At first, NLG computers generated text using templates. Like a game of Mad Libs, an NLG system would fill in the blanks based on some data or query. However, with the use of transformers, recurrent neural networks, and hidden Markov chains, natural language production systems have developed throughout time, allowing for more dynamic text synthesis in real time.

Similar to NLU, NLG applications must take into account morphological, lexical, syntactic, and semantic language norms while deciding how best to phrase replies. They address this in three phases:

- **Text planning:** In this phase, broad content is developed and arranged logically.
- **Sentence planning:** This step breaks down the information into paragraphs and sentences, takes punctuation and text flow into account, and, where necessary, adds pronouns or conjunctions.
- **Realization:** This step takes into consideration grammatical correctness, making sure that conjugation and punctuation standards are adhered to. For instance, the verb run is in the past tense as ran rather than runned.

How does NLG work?

Seeing a machine produce words out of thin air may feel like magic. In reality, though, it's just a few intriguing AI technologies collaborating. Among them are:

- **Language Model:** This AI's brain has been educated on vast volumes of text that you enter your prompt or data into. Language models can produce writing that sounds human by picking up on subtleties and patterns.
- **Natural Language Processing (NLP):** Consider NLP as the "reading" capability of the machine. The capacity of the machine to deconstruct and comprehend your instructions, prompts, and the data you supply is known as natural language processing (NLP).
- **Natural Language Understanding (NLU):** The main focus of NLU is "comprehension." It indicates that the computer examines the relationships and meanings in the data to guarantee that the text it produces is correct and logical.

These elements are used by NLG to create text methodically:

- **Data Input:** Prompts, spreadsheets, and database entries are examples of structured data that are sent to the NLG system.
- **NLP Analysis:** NLP deconstructs the data, recognises speech patterns, and examines grammar.
- **NLU Interpretation**: NLU guides the text production process by identifying the linkages and meaning within the data.
- **Content Planning:** The NLG system determines the content to be included and the sentence and paragraph structures.
- **Text Generation:** The language model uses the information and understanding from NLP and NLU to create the final output, which is human-readable text.

What is the Future of Natural Language?

To construct a human language AI system to pass the Turing test, developers concentrate on key phrases. The following terms are present in the entire end-to-end process when represented mathematically: An NLP system is created by combining NLU and NLG.

- **NLU (Natural Language Understanding)**: It is responsible for all processes, including decisions and actions.
- **NLG (Natural Language Generation)**: To reply, it creates the human language text from the system's structured data.

Consider the following real-world scenario to have a better understanding of their application: you have a website where you are required to provide share market reports every day. You have to conduct research, gather text, write reports, and publish them on a website for this daily assignment. This takes a lot of time and is tedious. However, if NLP, NLU, and NLG are employed here (Kavlakoglu and Vaish 2020).

8.2   Multi-Modal Capabilities in AI Agents

Multimodal AI agents are comprehensible information processing systems capable of analyzing data of different types and structures. They differ from other AI models in that the latter are commonly restricted to one input type (text or images, etc.) while mixing data from multiple sources creates broader context, increased flexibility, and greater effectiveness of answers. It has the prospect of bringing vivid improvements to human-AI communications by making the interactions more lifelike.

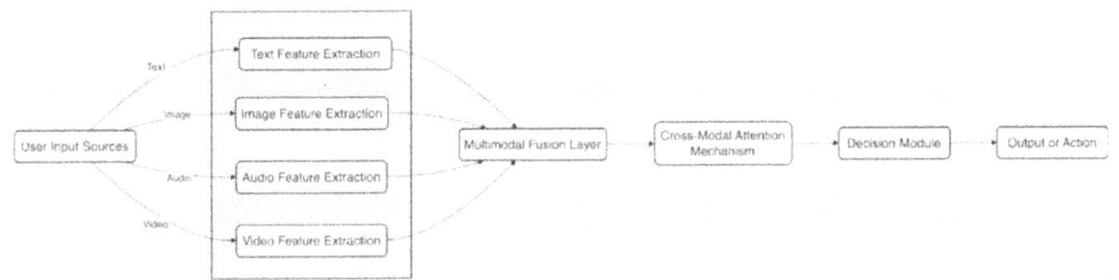

Figure 8.3: Multimodal AI Agent Architecture.

Source: – (Navdeep Singh Gill 2024)

The Core Architecture of Multimodal AI Agents

Developing effective multimodal AI agents involves integrating technologies and frameworks that handle distinct types of data inputs and processing.

Multimodal Fusion Techniques: At the heart of multimodal AI is the ability to merge information from different sources into a coherent representation. Fusion techniques can be categorized into three main types:

- **Early Fusion**: Combines raw data inputs at the initial stage before processing. This approach allows for rich joint feature extraction but can be computationally intensive.
- **Late Fusion**: separately processes each modality before combining the outcomes at the level of decision-making. It

is more modular but may miss out on deeper cross-modal interactions.

- **Hybrid Fusion**: Integrates features at multiple points, balancing the advantages of early and late fusion for optimal performance.

Multimodal AI agents learn from more than one domain of data, for instance, text, images, audio, and video. This capability is necessary where there is a need to navigate data with different types of data inputs, as well as legal requirements.

- **Visual Question Answering (VQA):** Using the features of image recognition and NLU to answer questions about visual content.
- **Speech-to-Text Systems:** converting spoken language to text for further processing.
- **Robotic Assistance:** Using aspects such as vision and touch as an interface to interact with the physical space.

Contemporary AI models like the GPT (Generative Pre-trained Transformers) utilize a deep learning architecture and provide record NLU and NLG efficiency. This is made possible by techniques such as fine-tuning on domain-specific datasets to improve the agent's ability to address special tasks. For instance:

- **Customer Support:** AI agents can respond to the user's question with empathy as well as accuracy.
- **Healthcare:** Virtual assistants can describe a medical treatment or a preliminary diagnosis.
- **Education:** An Intelligent Tutoring System is capable of answering students' questions and giving elaborate answers.

Key Applications of Multimodal AI Agents

- **Enhanced Virtual Assistants:** The existing virtual assistants, Siri and Alexa, can only respond to voice commands. These systems can be improved by adding Multimodal AI agents who can provide visual processing so that the system performs much better at handling queries that include images, face recognition, or gestures. This enhancement culminates in a more realistic and operational user interface experience. For instance, let us consider an application that can perform a voice command such as "What is it like outside?" In addition to understanding and responding according to the application's voice commandability, it can identify an object in a picture shared by the user and respond accordingly.

- **Healthcare Diagnostics:** In healthcare, each of the proposed multimodal AI agents can use data from medical images, patient records, and doctors' notes to generate diagnostic support. For example, an agent examining X-ray films and clinical text documents can assist medical personnel in diagnosis and treatment planning.

In addition, multimodal agents can be incorporated within telemedicine by augmenting the video consultation with continuous analysis of the patient's nonverbal cues, changes in voice tonality, and spoken contextual feedback.

- **Autonomous Vehicles:** The running of self-driving vehicles relies on the actual data from different sensors like cameras, LiDAR, and radar. Other data from multiple sources can help agents augment this information with traffic reports and GPS inputs to provide a solid decision-support system to prevent accidents and improve transport logistics.

Multimodal AI agents incorporate visual data from signs, pedestrians, and sounds, including sirens, honks, etc., and improve the situational awareness and decision-making abilities of autonomous vehicles.

- **Content Creation and Analysis:** It is also worth noting that the use of Multimodal AI agents is also transforming the generation and analysis of content. Those that contain bidirectional mapping for visual and textual data are employed in the automatic captioning of videos, interactive and multimedia narratives, and others. These capabilities integrate to optimize business processes in creative sectors and improve the experience for people with disabilities.
- **Education and E-Learning:** In education, using Agents increases the effectiveness and interactivity of education processes. For instance, agents can use text, images, videos, and audio to create rich lessons and tutorials. A multimodal tutor may have to teach the student about a particular idea verbally while illustrating the concept through diagrams and using verbal and visual or textual signals to answer the student's questions.

Challenges and Limitations

Despite their potential, multimodal models face several challenges:

- **Integration Challenges:** Combining data from diverse sources involves complex integration processes, which can lead to compatibility issues. Developing standardized protocols for data sharing and ensuring seamless interaction between different systems is crucial but often difficult to achieve.
- **Resource Intensity:** Training multimodal models often requires extensive computational power and memory. This resource intensity can lead to increased operational costs and may limit accessibility for smaller organizations or those with less technical infrastructure.
- **Interpreting Context:** Understanding context across different modalities can be complex, as each type of data may convey different meanings in various situations.

Developing algorithms that accurately interpret and respond to context requires ongoing research and refinement.

- **User Acceptance:** Adopting multimodal AI technologies may face resistance from users who are accustomed to traditional interfaces. Educating users about the benefits and functionality of these systems is essential to drive acceptance and integration into everyday use.
- **Maintenance and Updates:** Keeping multimodal AI systems up to date with evolving data sources and algorithms requires continuous maintenance. This can be resource-intensive and necessitates a dedicated team to ensure optimal performance and accuracy over time.

Future Trends: Multimodal AI Agents

1. **Integration of Multiple Data Sources:** Multimodal AI agents will utilize diverse data inputs, enabling more intelligent and context-aware interactions.
2. **Revolutionizing Industries:** These agents will transform sectors like digital assistants, diagnostic services, self-driving cars, and adaptive learning platforms.
3. **Overcoming Data Alignment Challenges:** As data alignment issues persist, advances in technology will lead to better synchronization of diverse data types.
4. **Addressing Computational and Ethical Challenges:** Ongoing work will address the heavy computational demands and ethical concerns surrounding the development of multimodal AI agents

8.3   Training for Industry-Specific Skills

While the deployment of AI adoption becomes integrated in various fields, educating agents in compliance with the knowledge of a certain field becomes crucial. Training methodologies include:

1. Supervised Learning: A machine learning method called supervised learning trains artificial intelligence algorithm models using labelled datasets to find the underlying patterns and connections between input characteristics and outputs. The learning process's objective is to develop a model that can accurately forecast results from fresh real-world data. An annotated dataset suitable for a specific industry, such as legal documents or medical documents, enables model training to become accurate.

Types of supervised learning

1. **Classification** Data is categorised using an algorithm in machine learning. It identifies certain entities in the collection and makes an effort to define or label those things appropriately. Support vector machines (SVM), decision trees, k-nearest neighbour, random forest, and linear classifiers are examples of common classification techniques. In difficult categorisation issues, neural networks perform exceptionally well. Neural networks are deep learning architectures that use layers of nodes that resemble the human brain to analyse training data.

2. **Regression** is employed to comprehend how dependent and independent variables relate to one another. Models try to predict the target output in regression issues when the output is a continuous value. Financial planning and sales revenue estimates are examples of regression jobs. Regression methods include linear regression, logistic regression, and polynomial regression.

Real-world supervised learning use cases

- **Image – and object-recognition:** Supervised learning techniques are helpful for computer vision and image analysis jobs because they can be used to find, separate, and classify objects from movies or pictures.

- **Spam detection:** Another illustration of a supervised learning model is spam detection. Organizations can successfully organize spam and non-spam correspondences by training databases to identify patterns or anomalies in fresh data using supervised classification algorithms.
- **Predictive analytics:** Predictive analytics systems are produced by supervised learning algorithms to offer insights. This enables businesses to predict outcomes based on an output variable and make data-driven decisions, which in turn helps executives defend their actions or change course for the organization's good.

2. Reinforcement Learning: A subfield of machine learning called reinforcement learning aims to teach an agent how to interact with its surroundings to learn the best course of action. Inspired by the concept of trial and error, RL algorithms learn through feedback mechanisms, where actions are rewarded or penalized based on their outcomes. Maximising cumulative rewards over time is the aim of reinforcement learning, leading to intelligent decision-making in dynamic and uncertain environments. Exposes agents to learning through experimentation within simulated environments.

Types of Reinforcement Learning

1. **Value-Based Reinforcement Learning:** Finding the ideal value function that gauges how desirable it is for an agent to remain in a particular state (or do a particular action) is the main goal of value-based reinforcement learning. Maximizing the value function, which stands for the long-term cumulative payoff, is the aim. Q-learning is the most often used method in this area. Q-Learning and Deep Q-Learning (DQN) are two varieties of value-based reinforcement learning.

2. **Policy-Based Reinforcement Learning:** Policy-based RL approaches, as opposed to value-based ones, seek to directly learn the optimum policy $\pi(a|s)$, which translates states into action selection probabilities. These techniques may work well in settings where value-based approaches are ineffective, such as high-dimensional or continuous action spaces. Reinforcement learning based on policies: Proximal Policy Optimization (PPO) and Reinforce Algorithm.

3. **Model-Based Reinforcement Learning**: Model-oriented. To forecast future states and rewards, reinforcement learning (RL) presents an explicit model of the environment. Before engaging with the environment, the agent simulates various actions and their results using the model. This aids the agent in better action planning. Model-based reinforcement learning types: World Models and Model Predictive Control (MPC)

4. **Hybrid Approaches: Actor-Critic Methods:** The finest features of value-based and policy-based reinforcement learning are combined in actor-critic approaches. These methods maintain two components:
    1. **Actor:** Learns the policy $\pi(a|s)$.
    2. **Critic:** Evaluates the value function $V(s)$.

The actor decides the actions, while the critic provides feedback on how good the action was, helping to adjust the policy. A popular algorithm in this category is Advantage Actor-Critic (A2C), which improves efficiency by calculating the advantage function (a refined measure of action goodness) rather than the raw value. Types of hybrid approaches include Deep Deterministic Policy Gradient (DDPG).

Real-World Applications

1. **Content Generation:** AI-generated content has the potential to become more customised to each user's likes and preferences.

2. **Autonomous Vehicles:** Autonomous car AI systems can improve efficiency and safety by learning from actual driving experiences. An autonomous car's RL agent might modify its course in real time in response to different incentives, such as time, safety, or fuel efficiency.

3. **Conversational AI:** Virtual assistants and chatbots are very helpful in customer service since they can have more conversational and contextual discussions. Reinforcement learning may be used by chatbots to improve their conversational models by using user input and interaction history.

3. Transfer Learning: A machine learning approach known as "transfer learning" uses insights from one task or dataset to enhance model performance on a related task or dataset. One Stated differently, transfer learning enhances generalisation in a different context by applying knowledge acquired in one context. Industry-specific models are fine-tuned on raised and concise datasets, thereby recommending a shorter period and computational load. There are several uses for transfer learning, ranging from deep learning model training to data science regression problem solving. Indeed, considering the volume of data required to build deep neural networks, it is especially alluring for the latter.

Types of transfer learning:

There are three related transfer learning sub-settings or activities.

1. **Inductive transfer:** Regardless of any differences or similarities between the target and source domains (i.e., datasets), this occurs when the source and target tasks are distinct. When architectures pretrained for feature extraction on massive datasets are then used for additional training

on a particular objective, such as object identification, this might appear in computer vision models.

2. **Unsupervised learning:** Due to the differences between the source and target tasks, this is comparable to inductive transfer. However, labelling source and/or target data is common in inductive transfer. Unsupervised transfer learning is just that—unsupervised, which means that no data has been manually labeled. In contrast, inductive transfer is a form of supervised learning.

3. **Transudative transfer:** This happens when the datasets (or domains) are different, yet the source and target tasks are the same. More precisely, the target data is usually unlabeled, whereas the source data is usually labeled. Applying information acquired from completing a job on one data distribution to the same task on another data distribution is known as domain adaptation, and it is a type of transductive learning.

Transfer learning use cases:

Transfer learning has several real-world machine learning and artificial intelligence applications.

1. **Natural language processing:** Feature mismatch is a prominent problem that impacts transfer learning in NLP. Connotations can vary depending on the domain in which a feature is found (e.g., light denoting weight and optics).

2. **Computer vision:** There is a plethora of literature on convolutional neural network (CNN) transfer learning applications due to the challenges in obtaining enough manually labelled data for various computer vision tasks. One prominent instance is ResNet, a pretrained model architecture that exhibits enhanced performance in applications involving object identification and picture categorisation.

Applications Across Industries:

1. **Finance:** AI is used for identifying risks such as credit or market risks.

   - **Fraud detection:** Models detect suspicious spending behaviours, thus preventing fraudulent transactions.
   - **Automated trading:** There are market analyses and trade performed through AI, where robots can analyse and trade significantly faster than human beings.

2. **Healthcare:** Artificial Intelligence helps in diagnosis and can recognize diseases such as cancer or diabetes from an image or a lab report.

   - **Drug discovery:** These models estimate which compounds might give an effective new drug, thereby saving a lot of time as compared to conventional methods.
   - **Patient management:** Virtual health assistants make appointments, remind patients to take their medicine, and provide first-level healthcare consultation.

8.4   The Interplay of Language, Context, and Task Performance

Thus, AI agents' performance is highly dependent on their capability to understand language in its contextual terms, which connects it to the specific task. Key factors influencing this interplay include:

1. **Contextual Awareness**: Situational awareness is needed because agents must know how to react to specific circumstances. For example, the phrase "Can you phone this?" has different connotations when used in discussions of finance and sports in particular.

2. **Dynamic Adaptability:** The pattern of AI systems' answer should change according to context flips in real time, like a change of the user's tone or topic.
3. **Task-Specific Fine-Tuning**: It is always more accurate and relevant when a model is developed to fit a particular need or context.

Real-world implementations demonstrate the critical role of this interplay:

1. **Virtual Assistants**: Apple's Siri and Amazon's Alexa, as good as they are, incorporate contextual clues.
2. **Legal AI Tools:** When evaluating contracts or referring to the case law, one has to take the liberty of using terminology of the domain.
3. **Customer Relationship Management (CRM):** To do so, AI tools use past interactions to adjust responses.

Multiple Choice Question (MCQs)

1. Which of the following best describes Natural Language Understanding (NLU)?

    A.   The ability to generate human-like text

    B.   The ability to comprehend and derive meaning from human language

    C.   The ability to process numerical data efficiently

    D.   The ability to translate text between languages

2. What is the primary purpose of multi-modal AI capabilities?

    A.   To enhance computation speed in AI models

    B.   To process and understand information from multiple data sources, such as text, images, and audio

    C.   To improve memory retention in AI agents

    D.   To ensure ethical AI decision-making

3. Why is training for industry-specific skills important in AI agents?

    A.   To increase their general-purpose capabilities

    B.   To optimize their performance in specific professional domains

    C.   To reduce computational costs

    D.   To improve natural language generation capabilities

4. What is a significant challenge in developing multi-modal AI systems?

    A.   Achieving bias-free algorithms

    B.   Integrating and aligning different data types like text, images, and audio

    C.   Enhancing short-term memory retention

    D.   Reducing language generation errors

5. How does context influence an AI agent's task performance?

A. By limiting the range of possible outputs
B. By providing relevant information that guides decision-making
C. By improving the computational efficiency of algorithms
D. By reducing the need for training

6. Which technique is commonly used for enhancing natural language generation?

A. Transfer learning
B. Convolutional Neural Networks (CNNs)
C. Knowledge graph integration
D. Gradient boosting

7. In multi-modal AI, what role do attention mechanisms play?

A. They filter irrelevant data to improve focus on critical inputs.
B. They ensure the model generates more text than required.
C. They reduce computational complexity by ignoring inputs.
D. They only process visual data while ignoring text.

8. What is the primary goal of training AI for task-specific performance?

A. To develop a universally adaptable AI
B. To improve performance in specialized and practical applications
C. To minimize computational power requirements
D. To maximize general-purpose language understanding

9. Which of the following is a critical factor for enhancing task performance in AI agents?

    A. Pre-trained language models
    B. Understanding task-specific constraints and context
    C. Data augmentation techniques
    D. Increased hardware capabilities

10. What does "language grounding" refer to in AI systems?

    A. The ability to detect grammatical errors in text
    B. Connecting language inputs to real-world concepts and context
    C. Improving the computational speed of language models
    D. The use of hierarchical neural networks in language processing

Answer

1	2	3	4	5	6	7	8	9	10
B	B	B	B	B	A	B	B	B	B

PART III

Applications of Generative AI Agents

CHAPTER 9

Conversational Agents and Chatbots

A Natural language processing (NLP) system that automatically answer in human language are known as conversational agents. Typically used as chatbots and virtual or artificial intelligence (AI) assistants, conversational agents are the real-world application of computational linguistics. A conversational agent is a virtual agent that may be used to have natural language conversations with people. Like humans, it comprehends context and meaning and mimics human-to-human interaction.

It allows consumers to order meals and perform other things by interacting with them via chat, text, or voice on computers, phones, and other devices. These may be achieved by utilising technologies like conversation management, natural language processing (NLP), speech recognition, text-to-speech synthesis, and machine learning (ML) to interact with people across a range of media. Use cases of a conversational agent: customer service, information retrieval, and revenue optimization. Examples of conversational agents that are currently being used in this market: Iris: a conversational agent for data science tasks, Woebot: a mental Health App, Roof.ai: a Real estate conversational AI chatbot.

How does a conversational agent work?

Technologies such as NLU, semantic analysis, text creation, dialogue management, and dialogue state monitoring are all employed in the programs that conversational agents use. They can comprehend what you say and react accordingly as a result.

They do this by translating spoken language into machine code. We refer to this method as automated speech recognition (ASR). The components, names, and semantics (meaning) of words are then identified using NLU. Lastly, text creation uses the collected data to construct phrases.

A dialogue manager is also operating concurrently with the NLU algorithms to make sure the discourse continues where it left off. This procedure keeps the discussion on course and enables the deployment of conversational bots in a variety of contexts.

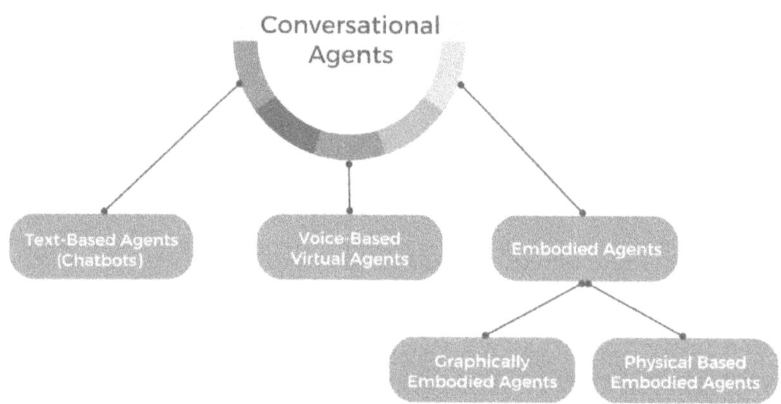

Figure 9.1: Example Types of Conversational Agents.

Source: – (dimension labs 2024)

One example of artificial intelligence that mimics human communication is chatbots. It may be used for tasks including completing projects, executing transactions, and responding to queries. If they are rule-based, they can only react by producing pre-written text strings or pressing buttons. There is no denying that these gadgets are goal-oriented. Customers may utilize them to help with online purchasing or to get specific answers to enquiries, including those about insurance.

9.1   The Evolution of Chatbots: From Scripts to AI

In order to simulate human interaction, the first chatbot, ELIZA, was created in the 1960s. It could identify keywords and respond based on pre-written scripts. Because these early chatbots were unable to comprehend context or learn from discussions, their interactions were restricted to rule-based ones.

Starting with simple written chatbots and progressing to the complex AI-powered ones of today, you can see the progression of chatbots in figure 9.2 below. The timeline tracks significant events beginning with the first chatbot, Eliza, in 1966 and continuing through the debut of PARRY in 1972, the growth of programmed chatbots such as Jabber wacky and ALICE in the 1990s and 2000s, and the arrival of virtual assistants such as Siri, Google Assistant, and Cortana in the 2010s. Advanced chatbots such as ChatGPT, Rasa, and Dialog flow appear in the 2020s, marking the end of the timeline and demonstrating how chatbot development has been driven by significant developments in artificial intelligence and natural language processing.

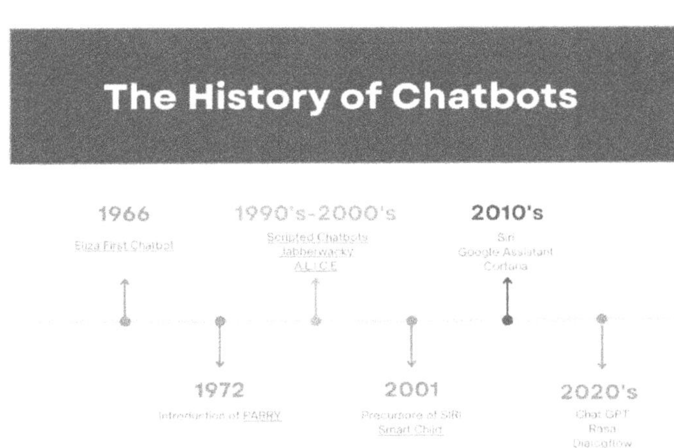

Figure 9.2: The History of Chatbots.

Source: − (Eliza 2023)

As technology advanced, chatbots went from being simple programmed systems to more sophisticated conversational agents powered by AI. Chatbots can now understand human speech and provide replies that sound natural because of developments in machine learning and natural language processing. This has made it possible for interactions to be more relevant and customized.

The advent of large language models, or LLMs, marked a turning point in this development. The use of LLMs trained on large datasets enabled chatbots to understand and generate language with remarkable accuracy, paving the way for more natural and meaningful interactions. Chatbots have evolved into versatile tools capable of handling complex tasks across several sectors, thanks to their evolution.

Progress in chatbot technology from early, more basic bots to state-of-the-art, conversational, and generative AI.

The Gist

- **Historical context.** An early chatbot, ELIZA (published in 1966), laid the groundwork for subsequent natural language systems by simulating conversation using pattern-matching.
- **Evolutionary path**. A long way has been covered in the fascinating history of chatbots, from simple rule-based systems to complex generative AI chatbots capable of natural-sounding chats and AI-powered personal assistants.
- **Challenges ahead.** The public prefers human customer service personnel over chatbots, and there are regulatory issues and the possibility of misinformation spreading due to AI chatbots.

ELIZA: At MIT, Joseph Weizenbaum created and introduced ELIZA, the first chatbot, in 1966. By employing model matching and replacement techniques, ELIZA was able to mimic speech and give the appearance of understanding. It would rewrite inputs as comments or questions to give some users the impression that they were speaking to a real person. The era of chatbots has begun, and they have developed to the point where they can now hold whole, existentialist discussions. From the first crude chatbots to the most sophisticated conversational and generative AI of today, let's look at the development of chatbots and how they are now being applied in a range of fields.

Rule-Based Chatbots: ELIZA served as one of the first crude chatbots. ELIZA was a conversation simulator that used simple natural language processing (NLP) and was named after the main character in George Bernard Shaw's Pygmalion. ELIZA's linguistic skills were mostly derived from individual "scripts." Open-ended questions and answers reminiscent of those of an empathic psychologist, such as Carl Rogers, were used to engage users in the most well-known script, DOCTOR. ELIZA might occasionally pass for human since she only knew basic pattern-matching rules and had no true comprehension of emotion.

PARRY: Jabberwacky was created in 1988 by British programmer Rollo Carpenter to simulate everyday human speech in a fun, humorous, and organic manner. The Loebner Prize, an annual AI competition to identify computer programs deemed most human-like, was awarded to it when it was made available online in 1997. Clever Bot, a chatbot that employed machine learning (ML) to converse with people, was introduced by Carpenter in 2008.

ALICE: Richard Wallace created A.L.I.C.E. in 1995, and it was the most human-like chatbot to win the Loebner Prize three times. It did not, however, pass the Turing test, which assesses whether a machine can have a conversation that is indistinguishable from that of a person.

Smarter Child: Smarter Child was a chatbot that AOL introduced in 2001 and that users could communicate with via AOL Instant Messenger. Smarter Child may play games, search for information, and engage in simple dialogue with users.

Personal Assistant Bots: The idea of conversational virtual assistants on phones was first introduced by Apple's Siri, which was debuted in 2011. Siri showed promise for voice-activated, AI-powered assistants that could comprehend commands in plain language. Two years after Amazon purchased the Polish voice synthesiser Ivona, Amazon Alexa debuted in 2015. It is now part of the Amazon Echo Dot, Echo Studio, Echo smart speaker, and Amazon Tap speakers.

Google Assistant for Android phones was released by Google in 2016. Microsoft created its assistant, Cortana, so it wouldn't fall behind. The introduction of these tech titans' virtual assistants signalled a shift in the acceptance of conversational AI as a standard and anticipated smartphone feature.

Conversational AI Chatbots: Initially, conversational chatbots were rule-based, which meant that they created replies using pre-written rules and templates. These days, conversational chatbots may generate better replies by using AI and neural networks. Replicas is an app developed by Eugenia Kuyda in 2017 that builds AI companions who can have tailored chats and gradually learn about their users. According to one story, Replicas is "the app that's trying to replicate you."

The chatbot's "Replika AI Agent" was trained on billions of lines of talk to identify subtle patterns in human speech. These ML techniques and neural networks are used by Replika to assess conversation settings and continuously enhance its answers depending on user ratings.

Generative AI Chatbots: Since OpenAI's ChatGPT was released in November 2022, generative AI has been in the headlines almost daily. Soon after ChatGPT was released, Google and Microsoft also unveiled their own generative AI chatbots, Microsoft Bing and Google Bard.

Generative AI has grown incredibly quickly and is now a major corporate commodity. The worldwide generative AI market is expected to reach $103.74 billion by 2030, up from $10.16 billion in 2022, according to Worldwide Data's Generative AI Growth Analysis. Additionally, because customers want speed and convenience when communicating with customer care, the usage of generative AI chatbots for customer support has increased dramatically. (Statista 2024).

Anthropic's Claude 2 is another well-known generative AI chatbot. Although Claude was developed as a personal assistant, it shares similarities with ChatGPT in that it is a broad language model. By using a method known as Constitutional AI, which seeks to imbue computers with "values" that are determined by a "constitution," Anthropic created Claude to be helpful, innocuous, and honest.

AI Chatbots Continue to Evolve: As AI chatbots continue to advance, programmers are discovering novel and distinctive uses for them. Embody Me's Xpression, an AI-powered camera software that lets users make any image come to life, is one fresh example of AI chat. With the help of this real-time generative AI tool, users may imitate their facial motions by using an image of their favourite historical figure, cartoon character, or celebrity. Xpression's developers hope that users will use the technology to sit in on business video conferences.

The SeamlessM4T chatbot, a multimodal and multilingual AI translation tool that mimics the Babel Fish from Douglas Adams' novel The Hitchhiker's Guide to the Galaxy, was recently unveiled by Meta. It has text-to-text translation, voice-to-speech translation, speech-to-text translation, and speech recognition for about 100 languages, and The Challenges of AI Chatbots Ext-to-speech translation. (Scott Clark 2023).

The Challenges of AI Chatbots

1. **Hallucination in Responses:** Generative AI chatbots sometimes produce inaccurate or fabricated information due to a lack of sufficient knowledge about a specific topic.
2. **Need for Guardrails**: Without safeguards, chatbots risk engaging in harmful or offensive discussions, potentially damaging a brand's reputation.
3. **Proliferation of Disinformation:** There are concerns about chatbots spreading incorrect or biased information, highlighting the need for robust safeguards to prevent disinformation.
4. **Transparency Issues:** Users often complain about not being informed that they are interacting with AI, prompting the need for brands to be transparent about their use of AI technologies.
5. **Compliance with Regulations:** Adapting to new and evolving AI regulations (e.g., EU AI Act, NYC Local Law 144) is critical to ensure responsible AI usage.
6. **Algorithmic Audits:** Thorough third-party algorithmic audits should be performed on generative AI systems to guarantee accuracy, fairness, and adherence to moral principles.
7. **AI Explainability:** The development and implementation of AI Explainability Statements are essential to provide transparency and accountability to the public, customers, and users.

As a result, talking about the so-called chatbots — at the initial stage, programs with a fixed dialogue script, and at the latter — AI technologies that allow having an intelligent conversation

— it is possible to state that the development of conversational technologies is very impressive. Modern chatbots use artificial intelligence and adapt to changes in the language processed, possess natural language processing, and use machine learning. Such a change has expanded the range of chatting bot use but also improved the performance of the tool in solving multifaceted tasks in the economy. As more and more advancements are witnessed in the use of AI in communication, it becomes clear that chatbots will take a central place in the transformation of the future of communication and service delivery.

9.2   Building Chatbots for Specific Industries

The general flexibility of chatbots means that they have been able to evolve to the numerous and dynamic requirements of the industries, thus revolutionizing these industries and the way stakeholder relations are handled. Chatbot functionality should be hired according to the industry to overcome challenges within sectors, enhance capabilities, and establish highly effective and satisfying experiences. In other industries such as e-commerce, banking, and health, among others, access to chatbots assists customers, manages financial queries in the banking sector, or giving health advice in the medical field, making industry-focused chatbots as critical tools in improving service delivery and operational performance. This section presents information on capturing specific industry needs and possible approaches and challenges facing the formation of chatbots for various industries.

The 6 Stages of Building an AI Chatbot

A well-defined methodology is essential for the creation of an efficient AI chatbot. Below is a summary of the six main things a Chief Technology Officer has to think about and do.

1. **Ideation & Requirements Gathering:**

   - **Focus:** Determine the main objectives, character traits, and scenarios in which the chatbot will be utilised. Find out how you will know they were effective, and make sure they align with your company's long-term objectives.
   - **CTO Role:** To comprehend customer demands and convert them into technological requirements, it is necessary to collaborate with business stakeholders. Scalability, security, and compatibility with current systems are important considerations.

2. **Design & Conversation Flow Architecture:**

   - **Focus:** Make a conversation flow diagram that can anticipate user queries and, by following their paths, get them where they want to go. Prioritise creating an interface that is both easy to use and consistent with your brand.
   - **CTO Role:** The fundamental design for managing conversation flows must be defined. Integration with a chatbot development platform or the use of application programming interfaces (APIs) may be necessary for this. Make sure that new features and integrations can be easily integrated into the architecture.

3. **Bot Building & NLP Implementation**

   - **Focus:** Create the chatbot's base functionality. Building complex chatbots from the ground up is one option, while leveraging existing models and APIs for natural language processing is another.
   - **CTO Role:** Select or build an NLP engine to let the chatbot understand user intent. Think about factors like language compatibility, accuracy, and the pros and cons of both custom models and pre-trained options.

4. **Testing & Training**

   - **Focus:** Train the chatbot with data from real or simulated interactions and do rigorous testing on its characteristics. Take care of any problems with its ability to comprehend and answer consumer enquiries.
   - **CTO Role:** Design a comprehensive testing plan that addresses all potential user scenarios, even those on the periphery. Constantly improve your methods by establishing routines to gather training data and retrain your models.

5. **Monitoring & Data Analysis**

   - **Focus:** Pay attention to how effectively the chatbot manages real-life conversations after it's enabled. Investigate user data for improvements and approaches to customise the discussion flow.
   - **CTO Role:** Create key performance indicators (KPIs) to monitor task completion rates, user engagement, and satisfaction. Use data analytics tools to improve the chatbot's algorithms and comprehend user behaviour.

6. **Optimisation & Continuous Improvement**

   - **Focus:** Continually enhance the chatbot's capabilities in response to user feedback and data analysis. This might mean improving conversation flows, retraining the NLP model, or introducing new features.
   - **CTO Role:** Make sure a feedback loop is incorporated into your chatbot development process. Create agile development techniques that enable quick iterations and enhancements based on actual data.

These methods will help you develop an AI chatbot that enhances your overall business strategy and offers a practical and efficient user experience, while also taking into consideration the specific duties of the CTO. (Janice Dombrowski 2024).

How AI chatbots can be used in different industries

In several businesses, AI chatbots have already emerged as effective instruments for improving the customer experience. These are a few examples of industrial use cases that are becoming more common:

Chatbots have been adopted across many fields to solve various problems and improve functioning in various organizations.

- **E-commerce:** Virtual shopping companions for customers involve helping with product selection, offering customers tailored suggestions, and overcoming the process of purchasing.
- **Banking and Finance:** Chatbots operate on a secure platform to respond to account enquiries, complete transactions, and even provide basic financial advising information.
- **Healthcare:** Chatbots driven by AI can organise appointments, answer frequently asked medical enquiries, rank patients according to symptoms, and even provide basic health advice.
- **Travel and Hospitality:** Chatbots assist with reservations for hotels, airlines, and activities; they also respond to questions about travel and offer suggestions for a more seamless trip.
- **Education:** Virtual teaching assistants and AI tutors customise learning experiences by attending to each student's needs and providing on-demand assistance.
- **B2B Chatbot:** Although chatbots are a new area of interest for most businesses, they are frequently underutilised in business-to-business (B2B) settings where interpersonal sales are a major focus. However, any B2B marketing

organisation may benefit greatly from having a 24-hour AI available to log enquiries and respond to basic enquiries.
- **Public Service / Community Chatbot:** Convenience and community service are not precisely the same thing, but chatbots can help alter that. Community chatbots may handle government communications with reporters, provide more details on press releases, and even assist in responding to important enquiries from the public.

However, the talk of building narrow, specialized chatbots for an industry needs a strong understanding of the needs and obstacles of the sector. Thus, organizations need to engage the domain-specific knowledge and incorporate the best technology options like Artificial Intelligence and Natural Language Processing that can help the organizations to build and develop the best of chatbot solutions that meet organizational needs efficiently. Chatbots have become more than just enablers of improving customer satisfaction; they are effective tools enabling operations optimization and innovation across vertical industries, becoming crucial in modern competitive environments. However, as they develop, generally, new opportunities in utilizing industry-based chatbots will arise allowing industries to progress even further. (One Beyond 2024).

9.3   AI Agents for Customer Support and Service

In the things related to customer support, the use of AI agents has played significant roles in rendering support at appropriate times and with high level of efficiency. Being able to carry out effective and repetitive questions, a chatbot relieves human operators to work on section requests that require more personal attendance, which in the long run enhances efficient service delivery with restrained response time.

Artificial intelligence (AI) customer service agents are smart devices that can comprehend and reply to consumer questions while staying within the parameters set by the agent. Agents are capable of providing individualised, conversational assistance with both basic and complicated problems, such as arranging a product return or responding to an often-requested query.

What are AI customer service agents?

Virtual assistants that can communicate with clients and aid with service operations are known as AI customer service agents. For a variety of activities, including multitasking and tackling complicated problems, they rely on machine learning and natural language processing (NLP). Most significantly, through self-learning, AI customer support representatives may consistently enhance their effectiveness.

Why are AI-powered customer service agents important?

There are several reasons why AI-powered customer support technology has become more popular:

- **Rising customer expectations:** 80% of consumers believe that a company's experience is just as significant as its goods and services, according to our study. As more rapid and individualised service becomes the standard, these expectations rise. Due to agents' round-the-clock availability, meeting these requests is quicker and simpler.
- **Increased desire for personalization:** Businesses must adapt to the demands of their customers for quicker, more individualized service. According to our research, 65% of consumers anticipate that businesses will adjust to their evolving requirements and tastes. AI can assess past consumer data, including preferences and behaviour, to

tailor customer service and even forecast future behaviour. AI customer support representatives may retrieve client data from a business's CRM to meet particular requirements while ensuring precision and uniformity throughout communications.

- **Shared knowledge:** Artificial Intelligence (AI) may leverage knowledge management systems to provide your customer support staff immediate access to pertinent data, facilitating quicker, more precise replies and raising customer satisfaction levels. Generative AI can even write articles for your team's knowledge base. This organisational information may be used by agents to create precise, conversational answers based on the knowledge base. For straightforward problems, 61% of consumers prefer self-service, according to our study, therefore, technology like agents may help businesses increase customer satisfaction while cutting expenses.

- **Rep burnout:** Additionally, according to our data, 56% of customer service representatives say they have burned out. Increased client needs and expectations, understaffed personnel, and monotonous tasks like repeatedly addressing the same enquiries are the causes of this. Agents may relieve your representatives of part of the burden by allowing them to concentrate on problems that call for empathy and human action.

Benefits of AI customer service agents

AI customer care representatives make sure that service teams always have adequate help to manage growing service channels and changing client demands. They may relieve human representatives of basic, repetitive duties and give them additional assistance to boost productivity and foster client happiness.

Here's a look at some of the benefits they provide:

- **Enhanced efficiency:** AI customer support agents' capacity to handle several client interactions at once significantly reduces response times and increases the efficiency of customer service operations. As a result, companies may now reply to more requests without compromising the calibre of their offerings.

- **Improved customer satisfaction:** AI customer service agents improve customer satisfaction (CSAT) by giving accurate and timely responses. They could enhance the overall client experience by customising interactions with data. Furthermore, they are committed to continuous improvement as they learn new things over time.

- **24/7 availability:** AI customer service agents are available around the clock to ensure that client enquiries are promptly answered. This continuous accessibility helps businesses meet self-service expectations and fosters customer loyalty.

- **Scalability:** Due to their ability to scale to accommodate increased client contacts, AI customer service agents are perfect for companies seeking to expand without sacrificing service quality. Consistent and dependable support is ensured by their easy adjustment to the increasing load as the number of instances rises.

- **Data-driven insights:** AI customer support representatives produce useful information on the interactions, preferences, and behaviours of their customers. Companies may utilise this information to learn more about the trends and wants of their customers.

- **Consistency:** AI customer support representatives consistently respond to consumer questions, fostering brand confidence and trust.

Those have shifted from rule-based systems to AI agents, which enhance contextual and individual approaches. Recent Chatbots enable the determination of user intent, the retention of context, and the ability to learn from past interactions, which increases customer satisfaction.

In addition, the application of voice recognition and natural language processing has added to the features available for a broad range of service scenarios within the AI agents. (Saleforce 2024).

9.4   Case Studies: Success Stories in Chatbots

Chatbots have become a game-changer in how enterprises engage with their customers and manage their processes and value propositions. Starting from the ability to engage each customer individually, ending with the possibility of automating all sorts of operations, chatbots are proving how effective they are. Organizations have reported a positive impact by adopting the use of chatbots in terms of customer satisfaction and organizational output, and in some cases, an increase in business revenue; hence, this technology remains crucial in managing current business operations. It's critical to understand that not every chatbot is a success.

The compilation of a successful chatbot examples list and scripts was compiled from eight applications to highlight several of the best and functional chatbots for different businesses. These are purely sales-oriented chatbots and rather friendly conversational bots programmed for enhanced customer interaction.

1. **GPT-4o:** It is difficult to discuss conversational AI without bringing up the most recent iteration of GPT. With many developers launching their GPT4-based chatbots, their inventiveness and ease of syntax make them an excellent chat buddy.

2. **Microsoft's Little Ice, XiaoIce in Chinese, became a social media phenomenon:** One of the most technically advanced chatbots on our list, along with GPT-3, is XiaoIce, Microsoft's largest success story. After launching in July 2014, XiaoIce had 0.5 billion talks in just three months. The company's valuation reached $1 billion after several rounds of funding, after it became independent of Microsoft in 2021 (Microsoft 2022).

   Successes: But what made the chatbot so successful? Since she's only proficient in Chinese, we couldn't do a full examination of all her potential. But these are the skills she talks about most: text mining, picture recognition, and context understanding, which allow her to speak fluidly and naturally.

3. **CustomGPT AI:** An artificial intelligence (AI) chatbot called ChatMTC was created by CustomGPT and the Massachusetts Institute of Technology (MIT) to connect entrepreneurs to MIT's entrepreneurship tools. The Martin Trust Centre for MIT Entrepreneurship aimed to consolidate knowledge from various sources into a single platform. They sought an AI solution capable of delivering accurate, citation-backed responses without fabricated information.

4. **Dr. A.I.? already makes recommendations like a doctor:** Developed by the healthcare app, Health Tap, Dr. A.I.? scans patients' complaints, asks follow-up questions, and analyzes their data saved in the Health Tap app to link patients with a specialized doctor in the neighborhood. "Dr. A.I." makes appropriate suggestions for the patient's next move by consulting Health Tap's database of responses penned by medical professionals. The latter can range from reading medical recommendations to booking a virtual or in-person doctor appointment or seeking urgent treatment.

5. **Woebot is a free therapy chatbot:** Woebot is designed by Alison Darcy, a clinical psychologist at Stanford University. It was released as an iOS app in 2018. Woebot generates pre-programmed responses for its users by utilising cognitive-behavioral therapy. 70 college students struggling with depression tested Woebot, and their changes were presented in a study paper revealing considerable advantages.

6. **Landbot.io for landing pages:** Businesses may create their chatbots with Landbot.io's visual, no-code builder. The chatbot built into Landbot employs conditional logic to turn leads into paying customers.

   A multinational insurance company has a 90% success rate in aiding existing clients with insurance claims and in turning prospects into paying customers.

7. **Domino's eases the process of ordering a pizza with a bot:** Domino's employs a restaurant chatbot to deliver a smooth order procedure. As part of their public relations campaign to raise brand recognition, Domino's created a chatbot for Facebook Messenger. With the bot, they are enabling customers to order pizza from any location. Customers can also use the bot to add a personal touch to their orders by specifying things like the type of crust they prefer and whether they want additional toppings (Cem Dilmegani 2024).

Today's business intelligence solutions use conversational agents and chatbots in a much bigger way than just handling customer questions. These technologies improve how organizations interact with customers by using smart language tools and learning programs. They provide users with personalized interactions that match their situations, making their experience better and helping people make good decisions. Now that industries are using digital change deeply, conversational agents will grow into new jobs by linking with IoT gadgets, learning what happens next, and managing real-time data updates. As these technologies advance, they help organizations be more effective, create new solutions, and win against competitors in our connected universe.

Therefore, this section has shown various success stories and thus acknowledged that chatbots have revolutionised many industries. They range from the use of chatbot technology to improve customer management, cut costs, and promote increased efficiency in the customer service organization, generate additional revenues to the organization, and optimize the distribution of services in the market. With future progress in artificial intelligence and natural language processing, the application of chatbots is set to increase substantially, allowing businesses to improve, grow, and provide great value to customers.

Multiple Choice Question (MCQs)

1. Which of the following best describes the evolution of chatbots?

- A. From rule-based systems to AI-powered agents
- B. From AI to hardcoded scripts
- C. From customer support to gaming applications
- D. From industry-specific tools to generic platforms

2. What is the primary limitation of rule-based chatbots?

- A. They are difficult to deploy in multiple languages
- B. They rely heavily on pre-defined scripts and lack flexibility
- C. They require advanced machine learning algorithms
- D. They cannot interact with customers in real-time

3. In which industry are chatbots most commonly used for automating repetitive queries?

- A. Healthcare
- B. Retail
- C. Customer Support
- D. Education

4. What is one advantage of AI-powered chatbots over traditional scripted chatbots?

- A. Lower deployment costs
- B. Ability to handle complex and dynamic conversations
- C. Minimal data requirements for training
- D. Easier integration with hardware systems

5. How do AI agents improve customer support services?

- A. By reducing the need for human interaction entirely
- B. By providing personalized, instant responses 24/7
- C. By replacing email-based support channels
- D. By focusing solely on premium customers

6. What is a key factor in building chatbots for specific industries?

    A. Incorporating domain-specific knowledge and terminology
    B. Using only open-source AI models
    C. Avoiding integration with existing systems
    D. Ensuring the chatbot remains generic for scalability

7. Which of the following is a challenge in deploying chatbots in healthcare?

    A. Ensuring patient privacy and data security
    B. Training chatbots to handle billing-related queries
    C. Managing chatbot interactions during peak hours
    D. Translating medical terminology into layman's terms

8. What distinguishes a successful chatbot case study?

    A. Demonstrated measurable improvements in user engagement and satisfaction
    B. Deployment in multiple languages simultaneously
    C. Reduction of employee workload by at least 50%
    D. Full automation of all customer interactions

9. Which case study is most likely to highlight the importance of chatbot customization?

    A. A retail chatbot offering personalized product recommendations
    B. A banking chatbot using pre-defined answers for FAQs
    C. A generic chatbot handling support tickets
    D. A chatbot for gaming with pre-programmed dialogues

10. What is a critical component of AI agents in conversational interfaces?

A. Natural Language Understanding (NLU)
B. Pre-defined response scripts
C. Integration with email servers
D. Static conversation trees

Answer

1	2	3	4	5	6	7	8	9	10
A	B	C	B	B	A	A	A	A	A

CHAPTER 10

Creative Agents

10.1   AI am in Content Creation

Content generation means producing useful content that matches your target audience's needs and keeps them connected. The digital era depends on content as its core channel to promote products, educate people, and share knowledge. People used to create content through manual talent and resource investment. Artificial intelligence systems are helping human content creators work faster and better without taking over their jobs.

Assembling content through an AI system brings important advantages to our technology:

a. **Rapid ideation and research:** AI systems can quickly generate concepts and locate pertinent information by analyzing vast databases.

b. **Scalability:** AI makes it possible to produce vast amounts of material without requiring a corresponding rise in human resources.

c. **Consistency:** AI may remain consistent across different types of content after it has been taught on a brand's voice and style.

d. **Personalization:** AI can scale up the customization of content to each user's preferences and actions.

e. **Multi-lingual capabilities:** AI can provide multilingual material, expanding international markets.

f. **24/7 productivity:** AI writers can produce material continuously since they don't sleep as human authors do.

Benefits of AI Agents for Content Generation

Before artificial intelligence (AI) bots started creating content, we depended on a combination of human ingenuity, research abilities, and frequently, simple willpower. Hours would be spent by content teams coming up with concepts, doing in-depth research, and creating each item from the ground up.(Ai 2024). In addition to being time-consuming, this procedure was constrained by human limitations in terms of information processing and creative output.

Creating content traditionally frequently entailed:

- Manual keyword research and topic ideation.
- Lengthy writing sessions with frequent writer's block.
- Multiple rounds of editing and proofreading.
- Struggles with maintaining a consistent voice across different pieces.
- Limited ability to produce content at scale.

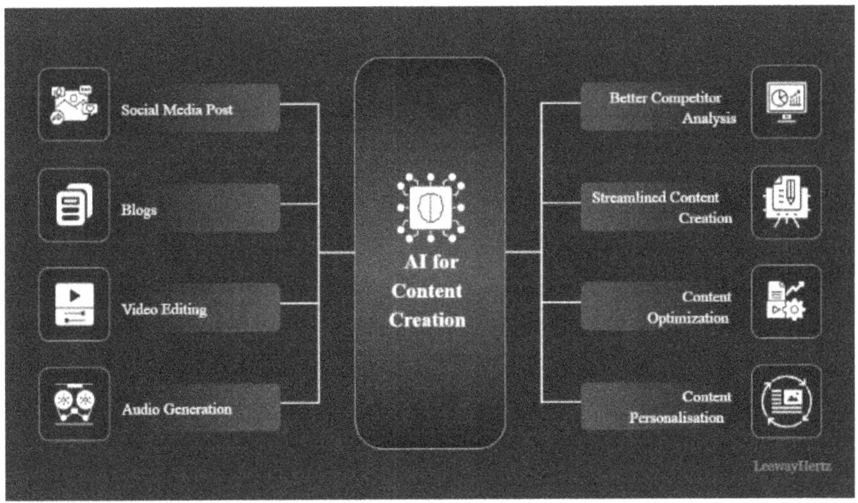

Figure 10.1: AI for content creation.

Source: – (Takyar 2023)

What are the benefits of AI Agents?

In ways that would have looked like science fiction only a few years ago, AI agents are revolutionizing the creation of content. They are enhancing human creativity rather than displacing it, freeing up content producers to concentrate on strategic planning and subtle narrative while delegating the laborious tasks of preliminary research and drafting. (Singh 2024).

Key benefits include:

- **Rapid ideation and research:** It would take people hours or days to generate subject ideas and acquire pertinent data, while AI bots can absorb enormous volumes of information in seconds.
- **Overcoming writer's block:** AI agents assist authors in overcoming creative obstacles and sustaining productivity by offering insightful prompts and recommendations.
- **Scalability:** Businesses are now able to satisfy the always-increasing demand for fresh information since content teams can now generate far more without requiring a corresponding increase in human resources.
- **Consistency in voice and style:** AI agents may remain consistent across different pieces of content once they have been taught on a brand›s tone and style, even when many human writers are engaged.
- **Data-driven optimization:** In order to continually improve content strategies and make sure that every piece resonates with the intended audience, AI agents may examine performance indicators and user interaction data.
- **Multilingual capabilities:** By using AI agents to generate content, businesses can expand their markets across languages at a lower cost through reduced translation expenses.

- **24/7 productivity:** Unlike real writers, AI systems never need to take breaks to rest. They operate without sleep to produce content in real time, so your product reaches customers sooner than ever.

AI tools mated to content generation processes now help humans and artificial intelligence work together to create even better content. This partnership between humans and machines impacts content production in ways that go beyond small improvements. Together, they transform what content creators can achieve. Our digital assistants will perform increasingly advanced tasks that expand content generation options beyond their current capabilities.

Potential Use Cases of AI Agents for Content Generation

Processes:

AI tools help writers and content producers solve their creation challenges in every business sector. These digital teammates function as more than content creation tools since they drive groundbreaking results in digital content work.

The most powerful potential application focuses on idea generation and group discussions. AI systems process whole sets of user data to recommend creative content solutions that appeal to their audience segments. They work as assistance tools for writers who use them as starting points to investigate new creative paths.

When it comes to content enhancement tasks, AI effectively produces results. These tools examine content for improvement opportunities, then suggest updates that fit SEO standards while tracking how audience members interact with the material. The repeated cycle lets content writers make changes that help their work survive the digital environment's constant updates.

Tasks:

Creation of content. When AI agents do particular jobs, their usefulness is evident. Consider the formation of headlines. Digital assistants can quickly generate a large number of impactful headlines while catering to certain channels and target audiences. This system's rapid headline creation enables us to do A/B testing to determine which captions produce the best results the quickest.

Present-day AI technologies provide users with useful initial drafts to build from. All content production tasks are initiated by these systems from preliminary drafts, which operate as a foundation for the work of human authors. The creativity process is both empowered and humanized when AI and human authors work together.

These algorithms are excellent at adapting written content to various cultural markets. By modifying already-existing information, they cater to a variety of markets and make helpful modifications according to regional languages and traditions. Businesses may use fewer translation workers to extend their content globally, thanks to the technology.

Although AI content tools have a lot of potential, they are better used in conjunction with human interaction rather than as a whole substitute. The greatest outcomes will come from content initiatives that successfully combine AI with human features.

10.2   Storytelling and Gaming Applications

Studies show that interactive content mixes gameplay with narratives openly transform both amusement and instruction parts. Through advanced AI technology these solution creators generate content from scratch while tailoring their design output to match user personas. Such capabilities not only enrich the similarly named scope of the individual, but also transform our approach to interaction and imagination in computerized realms.

Narration in Creative Agents

This project uses storytelling as a means of sharing cultural values, passing information, entertaining, and establishing knowledge. When it comes to the creative agents, the storytelling becomes a process that highlights its dynamic or interactive perspective. Through the use of AI, agents can create well-thought-out and culturally tuned storytelling pertinent to the specific inputs, patterns, and interactions of a user. These agents synthesize narratives by maximizing the application of NLP, machine learning, and knowledge graphs to create logically coherent narratives. (Unclear 2023).

For example, creative agents in interactive fiction platforms are capable of developing multiple paths that are formed as sub-trees with decision points that depend on the performance of the interactive agents. This offers agency and personalization, placing users into a narrative environment where they have an impact. This flexibility of achieving storytelling through creative agents makes this approach not just useful for gaming, but can also be implemented for education and therapy since most kids lean on narratives to learn better.

Gaming Applications

The gaming industry as a whole has been one of the first to embrace the use of AI-created creative agents to improve gameplay and design. They are now crucial for the creation of procedural content, for the works of NPCs and the "plot" creation. In contrast to conventional game design where even individual objects must be developed, creative agents built using artificial intelligence can develop infinite and diverse game worlds and game situations at the press of a button.

This is not as expanded as the previous one, but still one of the most stimulating – the generation of new stories for which the plot is traced depending on the actions of the player. This is done by the use of sophisticated programs which map players and base the game play and factors affecting the game in relation to the players. The creative game can engage creative agents to add new tasks or change the character relationships in response to how a player approaches the game, thus providing an individual approach.

In addition, the new AI-driven creative agents also improve the multiplayer gaming by further creating intelligent NPCs. Such objects can gain experience and change their behaviour in response to the player's actions – which is a lot more entertaining. The incorporation of stories into these systems guarantees that even those of the procedural or sandbox genre, can cope with the intrigue, that is erasing the difference between game design and storytelling.

Education Through a Game of Narrative

Now the usage of such applications as storytelling and gaming is not limited to entertainment only, it affects education and training. Instead of seeing learning as a boring concept, gamified storytelling, facilitated by various creative agents, makes the process look more like a game. Narrative aspects can be one of the means of addressing concepts that are rather abstract or difficult for the child to perceive.

For example, one creative agent could create a simulation game based on history, and in this game, people have to be certain historical personalities and make historical decisions. Such experiences develop critical thinking, increase tolerance, and an improve understanding of history. Similarly, in the corporate world, a game-based approach can also recreate practical business challenges whereby learners engage in solving business issues and decision making without real-life consequences.

Alterations and Limitations

As compelling as creative agents are on stories and games, there are issues and concerns with their application. A consideration that may rise is the level at which creativity competes with automation. While AI writing is fast, there's a danger of standardization of narrative and game design elements that may lose the subtlety and richness of work created by a human mind.

One of the most crucial challenges is an aspect of data privacy. The target users' data is an important concept for creative agencies since it is the core of personalization; however, there are critical questions regarding data collection, storage, and utilization. Transparency and users' consent remain vital aspects of creating trust, owing to privacy enhancement,(Morey 2015).

Furthermore, with manipulative narratives, which means that the AI agent uses psychological heuristics to control people's actions, an ethical review is needed. The increased role of creative agents renders the guidelines necessary so that agents operate within ethical boundaries and adhere to user agency.

Future Directions

The future of storytelling and gaming through creative agents can be seen in improving the processes of cooperation between people and robots. This movie concept shows that the marriage between human input and the processing power of AI can yield a generation of richer and diverse sensations. That is why such tools as AI that enable creators, for example, game designers, to co-design with machines and apply it during the game creation process help to optimize such work while maintaining the creative idea.

One of the emerging possibilities is the symbiosis of VR/AR and AI as its storytelling application. The integration of immersive technologies with adaptive narratives provides users with unique gaming experiences that may take the players into a completely different world. This has further implications for entertainment as well as for such areas as healthcare, where immersive narratives might be utilized in therapy or rehabilitation.

10.3   Creative AI in Art, Music, and Literature

AI and Creativity its Intersection in Art, music, and literature are all experiencing a big change thanks to AI. It's blending with creativity, creating new ways of doing things. This mix is bringing fresh ideas and pushing what we thought was not possible in art, music, and writing. In this blog post, we'll look into Intersection of AI and Creativity,(Blog 2023). Further, let's learn how AI algorithms are reshaping the creative process and pushing the boundaries of human expression.

AI and Creativity in Art

Figure 10.2: When fantasy becomes reality, it happening with two faces inspired by AI.

Source: – (Blog 2023)

1. **Generative Adversarial Networks (GANs):** Generative adversarial networks (GANs) have revolutionized AI-generated art. By setting two neural networks against each other, GANs produce original and visually stunning artworks. Moreover, artists like Mario Klingemann have leveraged GANs in their creative process. Also, they create captivating pieces that blur the lines between human and machine creativity.

2. **Deep Dream and Neural Style Transfer:** Neural networks are used in the Deep Dream and Neural Style Transfer procedures to reinterpret visuals in inventive and fantastical ways. These methods allow artists to explore new visual aesthetics and challenge conventional notions of artistic expression.

3. **AI Art Tools:** Deep Dream, Deep Art, DALL-E, RunwayML, Artbreeder, GAN Paint Studio and DeepArt Effects are just a few examples of the many AI art tools and generators available today, each offering unique features and capabilities for artists and enthusiasts alike.

AI and Creativity in Music

Figure 10.3: The above image shows a touch of the archaic with a dash of creativity and resourcefulness as far as space is concerned, and stories start flying off the pages.

Source: – (Blog 2023)

1. **Neural Networks in Music Composition:** Neural networks are being employed to compose music across various genres and styles. Also, platforms like Amper Music utilize deep learning algorithms to generate custom music tracks tailored to specific moods, genres, and preferences, empowering musicians to explore new sonic landscapes.

2. **Collaborative Performance with AI:** Creativity and AI Live music performances are increasingly including their intersection, which allows performers to communicate with intelligent systems that react to their input in real time. Furthermore, projects like Google's Magenta aim to explore the potential of AI as a collaborator in improvisational music-making.

AI and Creativity in Literature

Figure 10.4: The above images show an augmented person is depicted scrutinizing a data globule, in this context of a futuristic library, indicative of human and machine interconnectivity of sophisticated technology with knowledge centers.

Source: – (Blog 2023)

1. **Natural Language Processing (NLP) in Storytelling:** Generally, Natural language processing (NLP) algorithms analyze vast corpora of text to generate coherent narratives and dialogue. Nonetheless, interactive storytelling platforms such as Botnik Studios harness AI to co-create humorous and engaging stories in collaboration with human writers.

2. **AI-Generated Poetry and Prose:** Most importantly, AI systems are able to generate prose and poetry that closely resemble the tone and style of works written by humans. Additionally, initiatives like OpenAI's GPT models have shown this potential by producing evocative poetry and gripping stories. In doing so, they blur the distinction between human and machine writing.

Paradigm Shift: The integration of AI and Creativity, its Intersection in Art, represents a paradigm shift in artistic expression, consequently opening up new avenues for exploration and innovation. Additionally, if AI develops further, it will have a significant influence on the arts, encouraging cooperation, experimentation, and the discovery of new forms of beauty and meaning. Moreover, by embracing the symbiotic relationship between AI and Creativity, its Intervence in Arts, we stand on the brink of a renaissance, where machine intelligence and human inventiveness meet to expand the realm of artistic potential.

10.4   The Role of AI in Collaborative Design

If you deal closely with technology that can be automated by artificial intelligence (AI), the name may be alarming and make you shudder. A computer never stops learning and creating. They never eat, sleep, or want human contact, which may already be true for certain designers, but I don't blame others for feeling afraid. Whether we like it or not, the design industry must adapt to the new AI era that is upon us. Let's face it, as AI becomes more prevalent in our lives, some of us are thrilled while others are terrified that it will replace us in our professions or stifle our creativity. (Wilson 2023a).

But there is a silver lining. AI has the potential to improve a designer's process and create new opportunities when applied properly. This article will discuss the value of human creativity, how AI can support you during the design process, and how to use AI to enhance your job. Designers may accept AI as a useful tool rather than view it as a danger if they are aware of its potential and constraints.

AI's Role in the Design Process

AI is not a novel concept. Indeed, since the first part of the 20th century, people have been experimenting with the concept of a thinking machine that might carry out operations that normally demand human intellect. The Dartmouth Conference, a summer research session examining neural networks, the theory of computability, and natural language processing and recognition, is where the phrase "artificial intelligence" was first used in 1956. The AI Jeopardy competitors don't really like IBM's Watson, but in succeeding decades, AI will accomplish enormous advances in computer capabilities with greater government financing through initiatives like the Defense Advanced Research Projects Agency (DARPA) and, more recently, IBM's Watson.

As discussed in Sci-Fi UI: What Three Spaceships Can Teach Us about the Future of User Interfaces, artificial intelligence is already starting to advance to a level that many people only believed was feasible in science fiction. This technology is not limited to data science and development anymore. What part does AI play in design, then? Think of it as the exoskeleton of a designer. An exoskeleton is a wearable piece of technology that helps individuals by giving them increased power and support so they can work more effectively. As with an exoskeleton, the designer is not replaced but rather improved and maintains command over the AI's reasoning.

How Designers Can Leverage AI

A cybernetically augmented individual is depicted analyzing a holographic data interface, set against the backdrop of a library, symbolizing the integration of advanced technology with knowledge systems.

- Researching user problems
- Ideating solutions
- Visual design
- Validating those solutions

One of the most useful abilities a designer may have throughout the research phase is the capacity to recognize patterns in vast amounts of data gathered using both qualitative and quantitative research techniques. Additionally, AI is quite proficient in this field. Large volumes of user data may be analyzed by AI to identify patterns that a human designer would have overlooked, giving them a better understanding of the data.(Tré Wilson 2023).

There are many tools available to help designers throughout the brainstorming and graphic stages of a project. Midjourney is one of the most well-liked. This application helps designers come up with concepts for a variety of tasks, including interface design, graphic design, typography, and photography. Let's experiment with it first. I gave it a straightforward instruction.

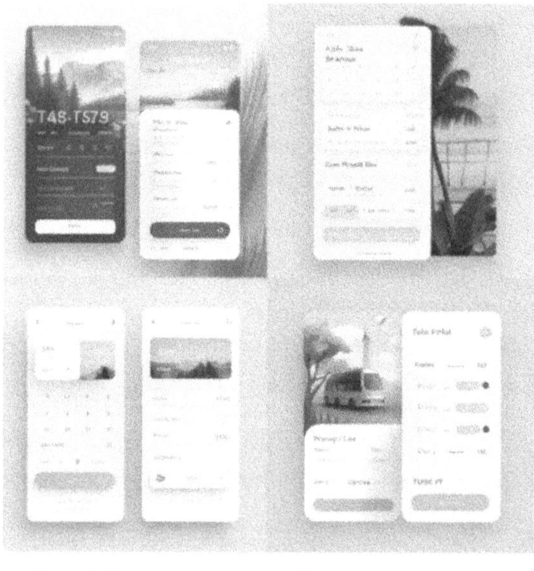

Prompt 1   Prompt 2

Figure 10.5: Difference between prompt 1 and prompt 2.

Source: – (Tré Wilson 2023)

Prompt 1: Design a user interface (UI) that is geared towards accessibility and trust for a user who recently joined our travel app.

Prompt 2: Make a user interface (UI) for a booking flow that takes into consideration the user's recent parent loss. Optimize for accessibility and trust for a person who has joined our travel app.

Mi journey's output is a fantastic place to start when brainstorming ideas for a usable user interface, even though it overuses images and shadows and lacks readable content. It's intriguing to observe how the user interface changed when a little adjustment was made to account for how a conscious human might feel. After entering that information, the layout appeared to have more flows and parts, featured less artwork, and lightened in colour, all of which suggested that the user was intended to swiftly and easily book their trip.

Booking a hotel for your parents' funeral is not a pleasant occasion, and the last thing you want is for the product to congratulate you on your purchase or say something like, "I hope you're excited for your trip!" Other times, product user experiences might be tone deaf or unnecessarily upbeat, as the majority of applications often,(Wilson 2023b). Would the user think the colour scheme and pictures were too upbeat for their circumstances? Given the wide range of scenarios a user may encounter, how neutral should an experience be? These questions will still need the designer to contextualize the information and take into consideration user flows, UX content and tone, micro interactions, and other factors.

The Importance of Human Creativity

AI is incapable of thinking or making decisions on its own. To stand out and be remembered, clients will rely on the creativity and inventiveness of a human designer to produce experiences that transcend basic layouts, interactions, and font selections. (Tré Wilson 2023).

Artificial intelligence poses a danger to designers due to its scalability, output speed, and quick learning. The fact is that moving pixels across a screen with a lot of effort is not a suitable solution. Even while many designers take great satisfaction in creating visually appealing landing pages and app screens, it's difficult to overlook how similar these layouts and features are. An artificial intelligence (AI) can easily compile samples from the internet and produce a design that satisfies the majority of users. The cost of this is that if tools like Mid Journey and Galileo AI only look at pre-made examples, they increase the already rising risk of UI trends becoming more uniform, which raises the question of how a client can get a product that can give their users a customized and unique experience.

Although visual design is still an essential component of UX design, it is becoming a talent that may soon be automated due to the ease with which AI can combine UI kits and design systems with designers. In a world where machines can perform tasks more quickly and efficiently, what skills will designers need to remain relevant?

Multiple Choice Question (MCQs)

1. Which of the following is NOT a common application of AI in content creation?

   A. Writing news articles
   B. Generating video game characters
   C. Diagnosing medical conditions
   D. Composing music

2. What is the primary advantage of using AI in storytelling and gaming applications?

   A. Increased game file sizes
   B. Automated bug fixing in code
   C. Dynamic and adaptive storylines
   D. Simplified user interfaces

3. In art, music, and literature, Creative AI can assist by:

   A. Predicting weather patterns
   B. Generating unique designs, compositions, or narratives
   C. Managing supply chain logistics
   D. Automating manufacturing processes

4. Which of the following technologies is often used in AI-driven collaborative design?

   A. Blockchain
   B. Neural networks
   C. Quantum computing
   D. Edge computing

5. How does AI contribute to collaborative design processes?

    A. By eliminating the need for human designers
    B. By automating repetitive tasks and suggesting creative alternatives
    C. By enforcing strict design guidelines
    D. By standardizing designs globally

6. What is a significant challenge in applying AI to storytelling and gaming?

    A. The high cost of GPUs
    B. Maintaining narrative coherence in adaptive stories
    C. Ensuring game compatibility across devices
    D. Automating 3D modeling

7. Which of the following is a famous AI tool for generating art?

    A. ChatGPT
    B. Stable Diffusion
    C. TensorFlow
    D. Jupyter Notebook

8. AI-generated music is commonly used in which of the following scenarios?

    A. Live orchestral performances
    B. Background scores for movies and games
    C. Advanced physics simulations
    D. Clinical drug trials

9. In literature, Creative AI can assist writers by:

    A. Creating entire novels autonomously
    B. Suggesting plot ideas, character arcs, and stylistic improvements
    C. Enforcing grammatical rules rigidly
    D. Replacing publishers entirely

10. The role of AI in collaborative design primarily involves:

A. Taking over creative roles from humans
B. Acting as a supportive tool to enhance creativity and efficiency
C. Replacing CAD software entirely
D. Eliminating the need for prototyping

Answer

1	2	3	4	5	6	7	8	9	10
C	C	B	B	B	B	B	B	B	B

CHAPTER 11

Productivity and Automation Agents

11.1 Coding Assistants for Developers

The development of software frequently encounters obstacles that lower productivity results. They encounter problems when looking for prepared code segments, discovering errors, and maintaining an even coding approach. Fixing errors in code takes up too much of developers' time and proves hard to do. Software developers need to track and resolve problems that stop the program from working properly. Debugging requires expertise and takes away a lot of valuable time from standard software development operations and slows down developer performance. They need to waste time searching for common programming blocks even as they detect issues and keep their code organized. Developers deal with an intense and hard process when locating and fixing problems in their programming code. Software developers search for and repair problems that make systems work improperly. Professional skills and time passing while being unproductive make up this procedure. Developers benefit from Generative AI Code Assistants, which help locate and solve problems while simplifying their debugging work and boosting their overall work output. These systems combine live code recommendations with automatic code standards to help developers focus on their work better. The technology helps developers work faster by detecting issues and creating fewer debugging tasks while keeping teams connected. Code Assistants help developers fix bugs faster while making coding and working more productive. The technology delivers code suggestions instantly while monitoring programming standards to help solve these issues. It lets developers accomplish tasks faster while cutting down on debugging needs and helping their team work better together. (Dr. Jagreet Kaur Gill 2024).

Using artificial intelligence technology including natural language processing and machine learning Generative AI code assistants help developers write code better while making software development easier. Let's explore how these assistants operate and what they provide along with their useful features.

How do Generative AI Code Assistants work?

Generative AI code assistants learn advanced AI and machine learning to support developers during their coding work. The systems study many programming projects from different online repositories before supporting programmers by decoding programming code and understanding keywords. The system trains on programming data to give smart ideas and make code segments that match the developer's work environment and needs. During interaction with a developer, the AI assistant examines all code context, which includes programming elements. Based on user input, the service generates possible code choices and predicts the next steps in the coding process. The systems make performance upgrades and propose ways to improve both code quality and speed.

Generative AI Tool Work Process

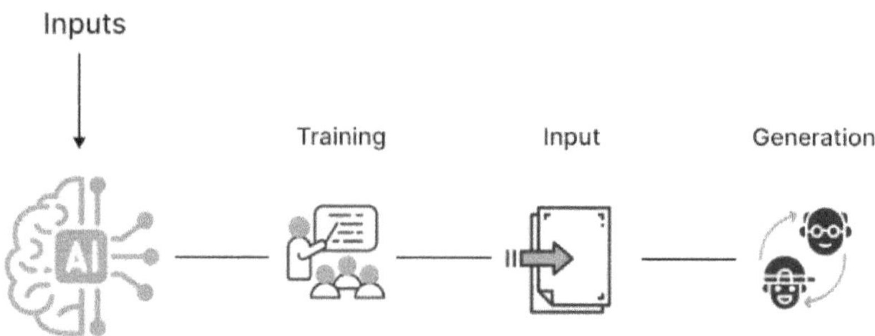

Figure 11.1: Generative AI Tool Work Process.

Source: – (Khambholja 2024)

This list shows popular generative AI code assistants that make development easier for developers:

- **Tab nine:** This leading AI-empowered code completion tool suggests smart programming code ideas across every programming language system.
- **Codota:** An Artificial Intelligence tool generates code suggestions that match project contexts and follow actual development patterns.
- **Kite:** An intelligent coding tool uses AI to complete coding tasks while showing code suggestions and references for better developer speed and output.
- **Deep Code:** A code review system uses Artificial Intelligence to find bugs and security problems in source code while assessing its overall quality.
- **Code Guru:** Amazon Web Services offers Code Guru, which uses artificial intelligence technology to review and enhance programming while boosting performance.
- **IntelliCode:** The AI-based code completion feature helps Visual Studio users write better code from within these two Microsoft development tools.

- **SourceAI:** Software development receives multiple benefits from an Artificial Intelligence tool that reviews and completes source code to enhance quality while making review tasks automated.

Through AI technology, these systems suggest smart code options to developers while completing their work and checking for code mistakes, so everyone can do more work in less time. (Dr. Jagreet Kaur Gill 2024).

Benefits of Using Generative AI Code Assistants

Machine learning coding assistants help developers work better and faster while writing code. Some of the key benefits include:

1. **Increased Efficiency:** Coding assistants complete routine programming tasks while freeing developers to handle complex design problems and innovative development tasks.
2. **Improved Code Quality:** Generative AI helpers generate better and shorter code from developers through recommended best practices in coding. Using this tool helps developers produce better code that is easy to understand and easy to maintain.
3. **Enhanced Productivity:** Code assistant tools make it possible for programmers to develop code at top speed through smart completion and creation options.
4. **Skill Development and Learning:** AI tools that generate code help users learn because they provide access to explanation texts and code examples from official documentation. These assistants help developers learn new skills as they work with them.
5. **Consistency and Collaboration:** When developers use these tools the team creates uniform coding practices because the system suggests accepted styles and rules. The tools make teamwork better by helping teams use the same coding methods across projects.

The generative AI code assistants help developers produce superior code results faster than before. These assistants use AI technology to help developers work more easily while transforming how software development happens today.

The development of generative AI presents us with many new ways to transform our world to significant progress. Generative AI applications work across different areas today, like various domains, including image generation, music composition, video game development, education and machine learning.

11.2   AI in Business Process Automation

What Is AI Business Process Automation?

AI business process automation uses artificial intelligence to enhance operational performance by automatically working through tasks and improving workflow effectiveness across company functions. Updated automation systems now benefit from cognitive technologies beyond basic preset rule enforcement. They acquire new knowledge to react directly to changing circumstances and conditions. Through natural language processing, they can speak with users while detecting multiple data types and creating new content. (Mah, Skalna, and Muzam 2022).

What Are the Key Components of AI Business Process Automation?

Using AI to automate business processes requires more than a single software application. Instead, it's a business transformation that incorporates several essential components that each work together to create intelligent, self-improving systems. These components include:

- **Machine Learning Algorithms:** Algorithms that enhance other capabilities like computer vision and big data analytics by analyzing historical data can recognize trends, forecast results, and make choices.

- **Natural Language Processing (NLP):** makes it possible for robots to comprehend, analyze, and react to human language. It is widely used in customer service automation for applications like chatbots. NLP is used in conjunction with robotic process automation (RPA) to automate workflows involving human communication, such as responding to emails or processing written requests.
- **Computer Vision:** The ability of computers to "see" the physical world. Often enhanced by machine learning, computer vision processes visual information and detects and categorizes objects. It can be integrated with IoT devices to monitor environments and processes in real time, feeding visual data into big data analytics platforms for further processing.
- **Robotic Process Automation (RPA):** Routine, rule-based automation is used to reduce the hours spent on tedious, repetitive tasks like paperwork. It works with other AI components like computer vision and NLP to automate more complex workflows.
- **Big Data Analytics:** Analyze enormous volumes of data to derive useful insights. The quality of AI automation solutions depends on the quality of the training data. Big data analytics ensures that high-quality, accurate, and real-time data is fed into AI automation so that it can provide advanced functionality, predict trends, and support decision-making.
- **Cloud Computing:** Lets you deploy and manage automation tools without investing in on-premises hardware. The cloud is a key component in AI automation because of its ability to process the large volumes of data necessary to train and deploy AI models for process automation.

- **Internet of Things (IoT) Integration:** Connects physical devices to the digital world, enabling automated control and monitoring of equipment and processes. IoT devices generate data that feeds into big data analytics and machine learning algorithms, enhancing predictive capabilities.

Why Is AI Important for Modern Business Automation?

AI is vital for modern business automation because it delivers unmatched efficiency, accuracy, and adaptability to organizational processes. (Boomi 2024). Here's why legacy automation technology just can't compete with the features and benefits of AI:

- **Enhanced Decision-Making:** Large volumes of data are analyzed by AI to produce actionable insights, which facilitate quicker and better business choices.
- **Improved Operational Efficiency:** AI automates processes 24/7 without fatigue, significantly reducing processing times and operational costs.
- **Predictive Maintenance:** AI detects equipment trouble so maintenance teams can make repairs before breakdowns happen, thus preventing unnecessary interruptions.
- **Personalized Customer Experiences:** Artificial intelligence systems use customer data to create personalized services that boost customer happiness and loyalty.
- **Streamlined Supply Chain Management:** By using AI to analyze data rapidly, you can better control your inventory and forecast sales, plus streamline delivery processes to run your supply chain better.
- **Automated Complex Tasks:** AI systems now take care of complex tasks that once needed human professionals, so users can focus on more complex work.
- **Optimizes Resource Allocation:** By studying resource use patterns, AI helps organizations find better ways to

distribute resources and eliminate unnecessary spending to make operations more efficient.

- **Accelerates Innovation:** AI's pattern detection in research data helps companies produce new items faster.

AI-Powered Automation vs. Traditional Automation

Before AI, automation was largely isolated to basic, routine tasks that followed a specific set of steps that could be programmatically defined. These solutions were an important step in the right direction, but had several limitations. Chiefly, legacy automation approaches struggle with any kind of variation. For example, a traditional OCR system can easily process invoices that are all in the same format with the same font and colors, etc. However, if a document comes with fields in different places, the wrong orientation, unusual fonts, foreign languages, or poor handwriting, the system often produces highly inaccurate results.(aws 2022).

Cognitive capacity is another important distinction between AI-powered automation and conventional automation. AI differs from traditional process automation systems due to its exceptional capacity for human-like thought, which includes real-time flexibility, autonomous decision-making, learning over time, and massive data processing power. These are all features that legacy automation can never replicate.

10 Ways AI Is Transforming Various Aspects of Business Automation

From customer service to manufacturing, AI is a transformational force. Let's look at some examples that highlight the breadth and depth of AI's impact on modern business operations:

1. **Intelligent Process Automation (IPA):** Through advanced automation algorithms and robotic process technology, IPA helps enterprises process complex business tasks such as compliance monitoring and financial audits faster.

2. **Advanced Data Analytics and Business Intelligence:** Data analytics and business intelligence now produce deep insights and predictions through artificial intelligence that help organizations make better strategic choices. Our company makes better decisions through real data insights to study customer patterns and spot future market changes.

3. **Autonomous Machinery and Robotics:** Modern machines and robots accomplish difficult work thanks to minimal human help in manufacturing and supply chains. Autonomous machinery with AI controls can perform advanced manufacturing tasks and automated logistics operations with minimal human involvement to boost performance.

4. **Natural Language Processing in Customer Service:** New chatbots using Natural Language Processing and generative Artificial Intelligence can serve customers on demand with instant support. Boomi boosts bot technology by linking chatbots to business systems, which now lets support staff serve customers with automated service anytime without needing human intervention.

5. **Predictive Maintenance in Industrial Settings:** Advanced cognitive platforms help industries detect equipment issues before they happen by examining sensor data. This allows manufacturers to prevent costly downtime and equipment repairs. Boomi is an excellent example of this. Our system connects IoT sensors with SAP to forecast machine issues and minimize equipment breakdowns.

6. **Personalized Marketing and Sales Automation:** AI creates better marketing campaigns while boosting sales performance by recommending personalized content to customers whose requirements AI silently examines.

7. **Supply Chain Optimization:** AI systems are transforming supply chain operations now. Advanced technology lets

companies better predict customer needs and optimize their stock flow for better supply chain performance.

8. **Fraud Detection and Risk Management:** Our ability to prevent cyberattacks against new threats depends on artificial intelligence. New AI systems for fraud protection screen transactions instantly to help reduce security risks.

9. **Human Resource Management:** New AI systems help HR teams select job candidates more efficiently plus they deliver tailored training to keep employees involved and predict how many staff members future operations will need.

10. **Financial Planning and Forecasting:** AI algorithms use large financial datasets to produce precise predictions, which help CFOs set better investment and budgeting strategies.

11.3   Tools for Team Collaboration and Workplace Productivity

Better team collaboration and better workplace routines assist project success in teams that use cloud-based BI systems. Business systems now enable groups to collaborate effectively while making work processes smoother. Companies use Microsoft Teams Slack and Google Workspace for easy communication and file sharing, plus they use tools like Jira, Trello, and Asana for project management. Cloud-based BI tools help teams share data in real time while creating interactive dashboards and examining data across departments. The tools help teams work together more effectively while overcoming department barriers to deliver better data-based decisions quickly.

Figure 11.2: The Best Generative AI Workplace Productivity Tools: Adobe Stock.

Source: – (Marr 2024)

Everyone has undoubtedly seen generative AI tools like ChatGPT and Mi Journey by now, which can produce computer code, drawings, and text that appear to have been created by people.

They're quite remarkable, but are you only playing with them, or have you figured out how to make them do anything really helpful?

Here is my compilation of what I consider to be some of the most remarkable, practical, or cutting-edge generative AI productivity solutions on the market right now. They can assist anyone in completing their work more quickly or effectively, freeing up valuable time for other fulfilling and imaginative pursuits. Although there is no set sequence for them, let's begin with five that I believe are very intriguing or significant:

- **Zapier**

  As an integration tool, Zapier facilitates communication and teamwork across various tools and apps. It may be used, for instance, to automatically record data from sources such as chat transcripts, emails, and phone logs. Additionally, it may be used to update data in CRM software like Salesforce or HubSpot, generate to-do cards in project management systems like Trello, and schedule events. Users can save a significant amount of time by doing this instead of copying and updating records on several platforms. Additionally, it lessens the possibility of human mistakes.

- **Trello**

Trello is one of the most widely used project management systems, used by millions of individuals to monitor ongoing work or gauge progress towards objectives. It has recently been enhanced with generative AI features, in this case from the Atlassian Intelligence engine, similar to many other industry-standard tools. This simplifies the process of making new cards by producing material and summaries automatically. It may even recommend changes to your writing for better language or clarity. Currently, exclusively accessible to premium and enterprise Trello customers, Atlassian Intelligence might eventually be incorporated into the free version.

- **Zoom with AI Companion**

During the COVID-19 outbreak, Zoom became a commonplace tool that let many businesses continue (nearly) as usual. Many others found it to be an excellent method to stay in touch with friends and family. The inclusion of Workplace Companion has more recently given it a generative AI makeover as well.

With a premium subscription, users may use this to generate automatic call and virtual meeting summaries, complete with assignments, follow-ups, and action items.

This allows participants to concentrate on the topic at hand and does away with the need to take notes and do other manual administrative tasks during meetings.

- **Be Done**

  In contrast to many of the productivity apps available online, Bee Done gamifies the task management process. By decomposing difficult activities into manageable parts and then giving rewards for finishing them, it does this. The concept is entertaining and offers a fresh take on the many "smart" time trackers, task managers, and to-do lists that are currently on the market. Users may compete against one another in "productivity tournaments" to determine who can become the most efficient, and Bee Done learns their behaviours to improve its capacity to optimize procedures.

More Great Generative AI Workplace Productivity Tools

Although there are many more tools available, these are some of the finest ones I've found that allow anyone to use generative AI to expedite daily activities:

- **Asana** automates workflow, calendar management, and project management duties.
- **Avoma.** In addition to transcribing sessions, an AI meeting assistant offers coaching insights to help users become more engaged and communicate more effectively.
- **Beautiful** uses automatic design features to turn concepts into fully realised slides and presentations.
- **Clara** is an AI assistant designed to function as a virtual employee by setting up meetings, facilitating communication between participants, and entering information into your calendar when it has been established.

- **Click Up** is a collection of generative AI-powered office productivity tools that includes whiteboarding, agendas, time monitoring, and project management.
- **Clockwise** is a time-management application that helps teams spend their time more effectively by automating scheduling.
- **Decktopus** creates presentations and slide decks based on natural language input.
- **Duet** is an additional Google service that Workspace has integrated to serve as a collaborative assistant for productivity activities.
- **Email Tree** Features of the email management and customer care platform include sentiment analysis, team management, CRM, multilingual assistance, automated reply ideas, and automation.
- **Fireflies** record, condenses, and looks up audio conversations.
- **Gamma** automatically uses generative AI to create presentations and other kinds of documents.
- **Mail butler** provides tools that automate writing, summarizing, and replying to emails by integrating with popular email services.
- **Notta** offers voice transcription and AI note-taking.

11.4   Personal Productivity Agents

An Overview of Personal Productivity Agents

Personal productivity agents are the breakthrough concept in the sphere of time and task management, individual, and even creativity. All of these AI tools are cognate to intending to smooth operations, speed up processes, and contribute to insights by automating tasks and offering optimized solutions. Unlike conventional where certain settings and input specifications are often fixed and must be coded, these agents employ state-of-the-art generative AI models to deliver interactive and contextual outputs. This consequently makes them an invaluable asset to any professional aiming at enhancing their productivity while at the same on strategic goals,(Bandi, Adapa, and Kuchi 2023) including their requirements, models, input–output formats, and evaluation metrics. The study addresses key research questions and presents comprehensive insights to guide researchers, developers, and practitioners in the field. Firstly, the requirements necessary for implementing generative AI systems are examined and categorized into three distinct categories: hardware, software, and user experience. Furthermore, the study explores the different types of generative AI models described in the literature by presenting a taxonomy based on architectural characteristics, such as variational autoencoders (VAEs).

The potential of Generative AI Productivity Agents

The prime competence of generative AI productivity agents is their flexibility. Such agents are apt at writing business emails, developing reports, even summarizing thick documents, or designing graphical content using basic commands. For instance, a marketing professional can employ the services of such an agent to develop an outline of a marketing campaign, or a project manager to take a brief of the week's activities. These tools are not simply for text-based exercises; they can create a chart or design a template, conscious or subconscious idea for a poster or an ad. These agents are thus always learning from the output and user feedback, making sure the output suits the style of the user.

Figure 11.3: Generative AI: A Productivity Powerhouse with Transformative Potential.

Source: – (Tm 2024)

Work Definition and Scheduling

As for the most significant benefit of the concept of personal productivity agents, one should mention the fact that these agents can exceptionally good at handling tasks exceptionally well. They provide an interoperable solution with calendars, task management, tools, and communication tools. They can interpret schedules, decide on conflicts, and prosaically suggest the best daily schedules. For example, if a user has back-to-back meetings, then the agent will recommend which particular time of the day can be dedicated to deep work or recommend that some tasks can be shifted to another day because they are not very significant. These tools can also send the required reminder and give information and updates at one go, and thereby cut stress. Such an approach relieves user of the stress that comes with multitasking and enables them to post their energies in a way that is harmonized with their working and social lives.

Role as Intelligent Advisors

Besides, making work activities efficient, personal productivity agents function similarly to knowledgeable assistants, using data analysis options. For instance, a business manager can employ an AI agent to analyze trends in sales, outlook of the market, and generate content for presentations that contains important figures, among others. These agents can also play out different acting scenes and present different decision-making to the users. They promote creativity in creative industries by offering new ideas, design solutions, or proposing and developing a prototype. This advisory function empowers users to discover choices, analyze threats, and execute fixes exponentially, making such agents essential for tactical planning.

Challenges in Adoption

Although personal productivity agents bring so many benefits, their implementation is not without difficulties. Privacy and security issues are crucial as operators of these agents very often deal with rather discrete or even classified data. Business and its stakeholders must guarantee that their information is stored and managed correctly within the ethics and the law. Besides, although such agents have the potential to make these decisions, they are not always accurate. They may, at times, generate results that are either misleading or prejudiced and therefore require scrutiny by the user. Nonsuch also causes overdependence and hinders learners from practicing critical thinking and problem-solving on their own, using such helpful tools as supplements to human reasoning.

Future Potential and Impact

The future of personal productivity agents looks very promising. In the future, with advancements in AI technologies, these agents are projected to further get smarter, able, and seamlessly embedded into operations. The first one is about how they can facilitate automation of trivial activities, improve creative activities, and act as smart assistants, which will allow professionals to do more within a shorter period. From healthcare to education and from creative arts to corporate management, these tools will revolutionize productivity measures. Incorporating artificial intelligence with advanced human endeavor, personal productivity agents will act as the cornerstone of realizing a future wherein technology and human capital will work hand in hand to define the future (Sihare 2023).

Multiple Choice Question (MCQs)

1. **Which of the following best describes the role of coding assistants for developers?**

    A. Automating manual coding tasks
    B. Assisting developers by providing code suggestions and debugging help
    C. Replacing developers in software development
    D. Writing entire software applications without human input

2. **How do AI-based business process automation tools benefit organizations?**

    A. By eliminating human employees
    B. By automating repetitive tasks and optimizing workflows
    C. By requiring more manual oversight and management
    D. By focusing only on the customer-facing aspects of business

3. **Which of the following is a key benefit of AI tools for team collaboration and workplace productivity?**

    A. Reducing the need for team meetings
    B. Automating all aspects of team decision-making
    C. Enhancing communication and task management across teams
    D. Eliminating the need for email communication

4. **Personal productivity agents are primarily designed to:**

    A. Replace human decision-making
    B. Automate repetitive personal tasks to save time
    C. Write and send emails automatically
    D. Perform physical tasks for individuals

5. Which of the following is an example of AI being used in business process automation?

    A. A personal assistant manages your daily schedule
    B. A chatbot answering customer queries on an e-commerce site
    C. An AI-driven tool automatically handles inventory and supply chain management
    D. A tool suggesting better ways to write code

6. Which of the following technologies is most commonly used by coding assistants to help developers?

    A. Machine learning models for pattern recognition
    B. Voice recognition software
    C. Virtual reality for immersive coding environments
    D. Data encryption for secure coding

7. In the context of team collaboration tools, AI can help with:

    A. Tracking employee performance in real time
    B. Scheduling meetings and prioritizing tasks
    C. Deciding on promotions and raises
    D. Replacing team leaders

8. How can AI tools improve workplace productivity?

    A. By eliminating human involvement in decision-making
    B. By automating manual processes and assisting with time management
    C. By providing workers with entertainment during breaks
    D. By controlling employee behavior through surveillance

9. Which of the following is an example of a personal productivity agent?

A. A project management tool that organizes tasks for a team
B. A smartphone app that reminds you to take breaks and manage your time
C. A chatbot that answers customer queries
D. An AI that performs payroll tasks for a company

10. Which is a primary challenge when implementing AI in business process automation?

A. High initial costs and complexity in integration
B. Lack of AI models for specific tasks
C. Increased risk of errors due to automation
D. Difficulty in automating customer support

Answer

1	2	3	4	5	6	7	8	9	10
B	C	B	B	C	A	B	B	B	A

CHAPTER 12

Autonomous Decision-Making Agents

12.1   AI Agents for Reasoning and Planning

Information retrieval, question answering, and even the creation and interpretation of art have all been transformed by ChatGPT and other generative artificial intelligence systems. However, those technologies are limited in their ability to do more intricate operations or connect several phases. The technology of generative AI is advanced by autonomous agents. To accomplish a single objective or a set of goals, they connect ideas and carry out several activities.

Overview of Autonomous AI Agents

In the field of generative artificial intelligence, autonomous agents are programs that use large language models (LLMs) to connect many ideas to produce a desired result or objective.

Autonomous AI agents differ from generative AI in that they may use tools and memory to do several tasks simultaneously without direct human assistance. The information stores that are sought and used in response to a prompt are represented by the tools that autonomous agents employ. These may consist of the LLM for the system or outside resources like databases, webpages, or other knowledge bases.(Shelf 2021).

The autonomous agent's learning experiences from previous prompts and outputs are referred to as memory. To complete the tasks at hand, these self-governing entities may access this memory and create more contextually appropriate replies.

7 Types of Autonomous AI Agents

Autonomous AI agents function with varying levels of complexity, from basic reactive behavior to advanced self-awareness. Understanding these different types will help you decide how AI can solve problems and make decisions on its own for your needs.

1. Simple Reflex Agents:

Simple reflex agents act solely based on the current environment, without memory or consideration of past actions. They respond to specific inputs with predefined rules. This makes them fast but limited in handling complex or dynamic tasks.

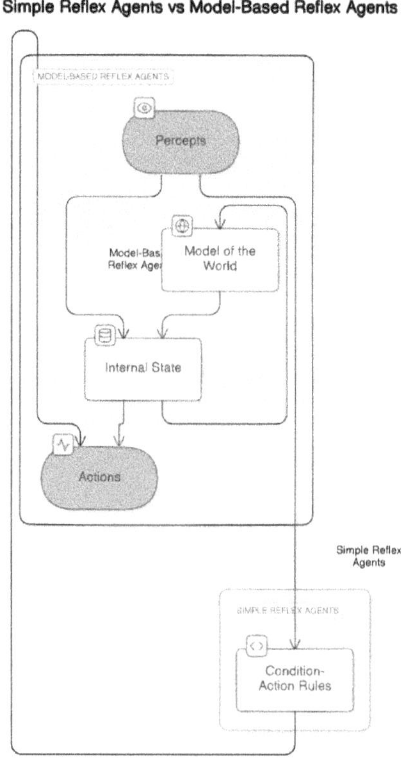

Figure 12.1: Simple reflex agents vs Model-based reflex agents.

Source: – (Jesse Anglen 2021)

2. Model-Based Agents:

Model-based agents use internal models to represent the environment so they can account for how actions will affect future states. This allows them to handle more complex scenarios by considering both present and future conditions.

3. Goal-Based Agents:

Goal-based agents go beyond reactive behavior by planning actions to achieve specific objectives. They evaluate different strategies to decide the best path forward. They are suitable for tasks that require decision-making aligned with long-term goals.

4. Utility-Based Agents:

A utility function, which gauges the worth or attractiveness of results, is used by utility-based agents to rank activities in order of priority. They assess which behaviors yield the best outcomes in addition to aiming to accomplish goals. The point is to balance trade-offs for optimal performance.

Figure 12.2: The above image represents a robotic man with artificial eyes, which illuminates: in his hands, there is an open book, which is a direct association with education. The image itself combines artificial intelligence with curiosity and is associated with futurism.

Source: – (Shelf 2021)

5. Learning Agents:

Learning agents improve their performance over time by adapting to new information and experiences. They use feedback from their actions to refine their behavior. They are highly effective in dynamic and unpredictable environments.

6. Hierarchical Agents:

Hierarchical agents divide tasks into subtasks and manage them across different levels of complexity. By breaking down problems into smaller components, they can handle intricate workflows and coordinate multiple operations simultaneously.

7. Multi-Agent Systems:

Multi-agent systems consist of multiple agents working together to solve problems or achieve common goals. The systems operate on documented communication patterns and good collaboration. These systems outstrip their ability to perform numerical processing and thus remedy complex problems that a single agent cannot solve independently.

Key Features of Autonomous AI Agents

AI agents come in many forms. There are AI agents from simple to complex, specific to broad coverage, and all of these are present at many levels of implementation. When looking for a business AI agent, you must take into account certain characteristics, which are relevant to your operating needs.

- **Autonomy:** An agent behaves autonomously if there is no need for human intervention at runtime during an operation. With development agents advancing in their development, they earn autonomy and are capable of operating without human intervention.
- **Adaptability:** Similar capabilities to what self-driving cars utilize to adjust their actions based on street conditions are needed by agents for adaptation purposes. An agent shows the ability to detect obstacles, process data and update its active patterns according to new opportunities through adaptive decision making.
- **Tool Use:** In order to do tasks through this system of AI agents, the AI agents need to have access to operate technology system application. Business agents conduct email marketing systems through automation software applications during their marketing campaign execution.

12.2 Optimizing Decision-Making in Complex Scenarios

Generative AI has evolved over the years to deliver the ability for autonomous systems to reason, learn, and make decisions across complex contexts. These agents operate in the conditions of knowledge uncertainty, high dimensionality of input data, and conflicting goals – all of which oppose such conventional deterministic paradigms as rule-based systems. In such conditions, achieving higher levels of decision-making optimality becomes crucial by using sophisticated approaches as well as promote flexibility in decision making while making sure that these systems are both very reliable,(Górriz et al. 2023).

The First Dimension of Complexity in Decision Making

Situations of this type are not easy for an agent to navigate when facing other agents, autonomous or otherwise. Such scenarios include those in which there are many linked variables and those in which the relationships are not linear, and frequently, external factors exert an influence that may well be unpredictable. For instance, autonomous driving is an environment where decisions have to be made rapidly based on dynamic stimuli that are characteristic of the environment where the self-driving car is situated; another example is the disaster response that is characterized by incomplete and noisy data, yet the consequences for a wrong decision can be huge. When the organization has more than one goal to be achieved, for instance, low cost but quality, or fast delivery but safety, it also adds another layer of confusion to the whole process. As a result, generative AI agents must be developed to operate efficiently in such environments.

Complex environments require that agents are placed where circumstances change within a short span of time. Large dimensionality brings in huge amounts of data with many attributes the handling of which needs complex mechanisms and data reduction. Furthermore, the reality of the world is usually rather vague and partially observed, so that stochastic models and learning strategies are required. Last of all, multiple objectives reveal the need to use multi-objective optimization paradigms to provide balanced and satisfactory results.

Strategies for Optimization

a. **Reinforcement Learning (RL):**

There is no question that reinforcement learning plays a critical role in programs that are committed to making decisions independently. It means that agents improving their performance in different settings by a series of trial-and-error interactions discover the best strategy at actual actions. Deep Q-Learning and Proximal Policy Optimization (PPO) are most relevant for controlling a high-dimensional action space and in cases with continuous decision making. RL lets the agent anticipate various results to select patterns that yield excellent long-term paybacks no matter the environment.

One of the most fundamental strengths in connection with RL is its flexibility. RL agents, on the other hand, increase their performance through the dynamics of the environment with which they operate. For example in robotics, RL based systems can effectively learn the task order and perform it from feedback to achieve higher efficiency with less error. As well as self-driving cars in which RL allows the creation of optimal control policies that depend on road traffic and environment condition.

b. **Multi-Agent Systems (MAS):**

When more than one agent is involved, decision-making process can be greatly improved if all of them work together. Multi-agent systems build in on cooperation to utilize the available resources as well as the greater number of views on the intended objectives which may not be achievable individually. Through coordination mechanisms like the auction-based systems and swarm intelligence, issues like efficiency and scalability arises, which is why MAS is more advantageous in application like logistics where multiple tasks have to be coordinated and distributed.

The integration of interaction into the multiple agents make the system more complicated because the agent has to negotiate and coordinate with the other agents. Recent developments in communication protocols and Distributed Computing systems have made it possible for agents to intercommunicate seamlessly. For instance, in smart grid by Pratyaksha Srungavarapu, MAS enhances energy management by adjusting the supply output of many nodes to meet a given demand. Moreover, in areas of search and rescue missions, MAS allows multiple groups of drones to work together and area can be surveyed faster than by one drone.

c. **Bayesian Decision Theory:**

Gricean approaches present a great way of using prototypes in Bayesian models with great effectiveness for appreciating uncertainty. Through encoding probabilistic relationships between variables, the BN and GP allow agents to modify the currently held knowledge based on fresh information. This is important in particular situations where conditions and data inputs are often changing.

Application of the use of Bayesian approaches is most preferred where there is little statistical data available or the data is noisy. For instance, in medical diagnostics, Bayesian models enrich the prior by the actual data of the patient in order to make a better prognosis. They used in the field of financial forecasting to include probabilities of fluctuation in the market in the course of proposing potential outcomes for investors to make informed decisions.

d. **Generative Models for Scenario Simulation:**

Realistic simulations require generative model where variational autoencoder VAEs and generative adversarial networks GANs are commonly used since they are able to generate realistic profiles. These models enable the agent to try out various strategies as a simulation is a replication of the real world with the added advantage of fostering a lower risk exposure when compared to live models. For instance, in the healthcare industry generative models can generate similar patient reactions to different treatments and the next steps are to improve and tailor these treatments.

Scenario simulation does not end at risk mitigation. Therefore, these models help expand the generalization abilities of the agent by presenting it with different hypothetical scenarios which the real world may pose to it. In urban design, generative models recreate trafficked routes and test how traffic would behave under different scenarios to enable policymakers to create effective transport systems. In the same way, in defence these models offer virtual scenarios for strategy formulation and threat assessment.

e. **Meta-Learning:**

Meta-learning or "learning to learn" ends with agents the capability or less adapt to the new task. There are methods that allow an agent to apply decision making proficiently across all conceivable scenarios, such as Model-Agnostic Meta-Learning (MAML). It is imperative today especially when addressing disaster response among other fields whereby the agents are required to perform their tasks in a novel way.

The future of meta-learning is based on the fact that it allows to shorten the training time and increase the work on new tasks. For instance, in e-commerce, meta-learning systems can easily fine-tune well to changes in customer preference and perform personalization with little extra training. In scientific research, these systems help to proceed with hypothesis testing by organizing comprehensive experimental space.

f. **Explainability and Interpretability:**

It is mandatory that some of the most important decisions should be made known to the public and the shareholders should be held accountable for how some of them were arrived at. SHAP and LIME methods convey information on what caused a specific action by an agent. These tools allow developers and stakeholders to both understand and verify the correctness of autonomous systems, which will lead to increased use of generative AI.

Regulatory, legal and accountability requirements as well as ethical considerations are the other areas that can be enhanced by explainability. The role of decisions has profound impacts that affect the society and thus there is a need to explain it using AI result. For instance, explainable AI in loan approval systems' policies guarantee that every criterion, such as income rate and credit record, is equally enlisted so that people can understand why they were rejected a loan.

Tools and Frameworks

There are many tools and frameworks for the optimization of decision making in self-contained agents. Reinforcement learning has been tested on OpenAI Gym providing it with a stable environment for developing and comparing learning algorithms. As for AI toolkit, Unity ML-Agents allows to train agents using deep reinforcement learning within simulated 3D environments, containing a lot of potential for complex testing. TensorFlow Probability and Pyro enable agents to work with decisions containing ambiguity through the application of probabilistic inference and Bayesian modelling.

Its Usage in Different Sectors

There are a number of industries that have benefited from the optimization of generative decision-making AI. In health care, the machines self-organize in decision-making processes involving treatment choice, risk, and cost, and patient characteristics. In the case of the operations management of the element of order, dynamic route planning of deliveries enhances efficiency while on the other side inventory management of spare parts should enhance efficiency by reducing the operational costs of the company. Financial applications are as follows: They include portfolio management systems that are AI implemented to allow real-time changes based on market trends. Pest control uses generative AI models whereby AI predicts pest outbreaks and determines the best approach in mitigating them without harm to the environment or business.

In manufacturing, autonomous agents facilitate the reduction of time as well as cutting costs and minimizing compromises to product quality. These systems are capable of providing real time alerts of the deviations which can cause expensive defects and downtime. In agriculture, generative AI is applied in precision farming through which it can detect fate, weather condition, and soils through which productivity can be boosted while using minimal resources. In the same way, in education, the use of artificial intelligence offers learners customized learning environment and offers them content according to their preferences.

Issues and Prospects

Albeit, there are still several issues of large importance which have to be solved in the frame of the generative AI's development. Of most importance are the ethical issues such as fairness, accountability and issues to do with privacy. Scalability is also another consideration whilst system must be resilient to attack and sudden failure. This becomes hard when optimization techniques are extended to accommodate larger and more complicated systems. Last but not the least; the incorporation of the autonomous agents into existing workflow and other infrastructure is complementary to achieve the optimal use of AI (Prem 2023).

Information for future research should explore a synergy of the examined optimization approaches as heuristic and metaheuristic methods. Appropriate human involvement with artificial intelligence offers an opportunity to get better results more efficiently by combining human cognition and machine substantiation. Specific adaptations for various industries will also push the development of generative AI to meet solutions to specific problems in particular fields. In these areas, all the more, autonomous decision-making agents can further on maintain the disruption across industries and deal with more sophisticated conditions (Ali et al. 2024).

The key to the broader implementation of these systems will be ethical frameworks and transparency standards integration. As we progress through the development of generative AI, the concern and focus for building ethically sound and equitable products will assume more significance. This guarantees that the capability of independent decision-making agents to bring about a change for the whole society is channeled positively.

12.3   Applications in Strategic Business Decisions

Generative AI agents have the potential to revolutionize operations and processes across a wide range of industries. These disruptive innovations have a great chance of changing how things are done and what is produced in the following market areas:

a. **Marketing and Advertising:** Using AI generative agents, businesses can now create tailored advertising campaigns and content that is fit for purpose and more likely to appeal to their target audience.

b. **Media and Entertainment:** Using scripting techniques, gaming effect development, and musical composition, media and entertainment platforms utilize generative AI agents to create a variety of material.

c. **E-commerce and Retail:** Generative AI agents, which include a mix of generative AI agents that suggest individualized product ideas and automated digital assistance features, enable customers to have better shopping experiences.

d. **Healthcare:** Generative AI agents that assess pictures can be used by medical practitioners for research of medication and patient treatment planning, as well as diagnosis support, among other healthcare applications. Generative AI synthetic outputs can be used for training data by machine learning models, thereby guaranteeing privacy compliance restrictions.

e. **Finance and Banking:** Generative AI agents in financial institutions make it possible to achieve two convergence points: fraud detection and risk assessment based on algorithmic trading and personalized financial recommendation services. Real-time financial data is examined by financial institutions in order to find anomalies and uncover crucial patterns in order to make better decision.

f. **Automotive and Manufacturing:** Production enhancement capabilities and prototype development capabilities as well as product performance simulations are offered by generative AI agents for manufacturing and automotive operations. Using the systems, we merge predictive maintenance capabilities and quality control support features into a single system.

g. **Gaming:** Through AI agents that automate environmental area and game level design for unique characters, a Video game procedural content emerges. Process-driven systems improve user experiences through dynamic mechanism design and adaptive storytelling.

h. **Education and Training:** Intelligent agents, under developer direction, build modified educational content, as well as interactive educational solutions and visual information. Adaptable learning solutions combined with generative AI allows personalized support programs to be delivered that provide timely feedback to users.

i. **Legal Services:** Through generative AI agents, legal service organizations can automate document generation and possess legal document and contract analysis capabilities. They support lawyers to access preparatory legal assistance alongside linked relevant precedents through their case reference solutions (Doanh et al. 2023).

Use cases & Benefits of Generative Agents for your business

Generative AI agents provide several operational benefits to a wide range of business industries. (Doanh et al. 2023). Several benefits for business operations are offered by these systems:

a. **Content Generation:** Generative AI agents enable businesses to maintain continuous digital relationship with their audience without spending a lot of time or effort to do that by using automatic algorithms to produce contents such as blog posts, product description, marketing material etc.

b. **Personalization:** Given the client's data, AI agents create personalized guidance. This guidance combines commercial communications with suggested fees that reflect both ordinary behavioural patterns and verified consumer preferences. They're essential for successful customer conversion because advanced customization technologies guarantee customer happiness.

c. **Creative Design:** Generative capabilities of AI agents make it possible for businesses to create new graphic designs, logos, development prototypes, and more. Generate Multiple Design Variations allows firms to speed up the design option examination process by using system defined criteria.

d. **Virtual Assistants and Chatbots:** Using generative AI agents, businesses can develop customer care bots to provide observational consulting for order monitoring and scheduling request, as well as advice suggestions to both consumers and service users. This system is efficient, offering higher quality services, but it still needs human interaction for non-routine operating requirements.

e. **Product Innovation and Prototyping:** Generative AI agents are used by businesses to create new business innovation concepts and product prototypes. Business innovation characteristics emerge from a strategic study that links primary market trend analysis to research on customer preferences and a review of current items.

f. **Data Analysis and Insights:** AI generative agents› data analysis through data-driven decision making to maximize business strategies provides data analysis insights that humans may overlook while manually examining. The process supports operations management by making data-driven decisions that can improve business operations and show the growth potential of the business.

g. **Risk Management and Compliance:** These artificial intelligence models evaluate official regulatory provisions to detect operational threats along with business noncompliance risks. Through generative AI tools businesses proactively discover risks while meeting regulations so they can protect themselves from potential liabilities.

12.4   The Role of AI in Healthcare Decision Making

AI healthcare agents are artificial intelligence systems that carry out specific tasks connected to healthcare, such as organizing appointments, generating treatment recommendations, analyzing patient data, detecting anomalies in medical images, and automating daily tasks. (Nawaz 2024).

Figure 12.3: AI Agents in Healthcare.

Source: – (Nawaz 2024)

To gather environmental data, think critically, and perform necessary activities on their own, agent programs use machine learning in conjunction with natural language processing, deep learning, generative AI, and computer vision.

<u>Applications of AI Agents in Healthcare</u>

AI-powered healthcare agents have the potential to transform several tasks and workflows.

a. **Medical Imaging Analysis:** To enhance medical image analysis, more than 75% of radiologists think about utilizing AI algorithms. AI agents with computer vision capabilities can help radiologists diagnose problems by analyzing medical pictures from CT, MRI, and X-ray scans. They can identify small abnormalities and encourage the early diagnosis of serious illnesses including artery blockages and cancer.

b. **Disease Diagnosis and Risk Prediction:** AI bots can help doctors diagnose diseases more accurately and make 30% fewer mistakes. These agents use databases to produce evidence-based therapy suggestions after receiving patient data inputs.

c. **Virtual Nursing Assistants:** Through these technologies, AI virtual assistants aid healthcare practitioners in providing precise remote patient care. With the help of virtual nurses, AI agents can become healthcare monitors that check patients' well-being and provide immediate primary medical support. With the use of this application, healthcare facilities in underdeveloped nations may use proactive patient care to cut hospital readmissions by 20%.

d. **Patient Education & Information Delivery:** In order to interact with patients, artificial intelligence systems are trained to use every day spoken language. Executives working at the front desk manage patient appointment alerts, report sharing, and appointment rescheduling while educating medical professionals about treatment protocols. These agents are ideal digital substitutes for human front desk employees.

Benefits of AI Agents in Healthcare

AI agents that increase operating speed while achieving better medical forecasting accuracy improve healthcare in a number of ways. A brief summary of the main advantages of AI healthcare agents is provided here, along with some real-world examples.

a. **Improved Diagnosis and Treatment:** AI agents can accurately interpret medical photos and reports thanks to sophisticated AI algorithms and computer vision. They are able to identify minute anomalies that the human eye could overlook. For the healthcare sector, extracting important medical information from complicated and noisy pictures has become feasible.

b. **Personalized Service Delivery:** Healthcare providers may utilize AI agents to develop customized treatment programs in a world where customization can cut client acquisition expenses by 50%. AI systems can analyze past treatment data and recommend 100% personalized healthcare facilities and preventative care programs.

c. **Enhanced Drug Discovery and Development:** Healthcare researchers can swiftly find the best medication prospects because to AI health aides' ability to mimic modular interactions. By determining the best patient groups and forecasting precise results, they may assist this sector in optimizing clinical studies. As a result, the medical business advances through faster yet more efficient medication discovery.

d. **Improved Patient Care:** Every year, 5 million people are killed by inadequate healthcare. The main obstacles to providing high-quality services are a lack of resources, sufficient patient data, and early detection, all of which AI agents can easily overcome. AI agents can stay available to monitor patient data in real time and instantly forward it to the appropriate department. They can expedite illness identification by analyzing patient data as it is received.

AI agents can assist medical practitioners in identifying the corresponding disease as AI models are able to link two distinct sets of data. They can give a more comprehensive perspective for thorough illness diagnosis by analyzing many data types at once, including imaging, genetic information, test findings, patient history, etc.

The power of autonomous decision-making agents has led to the growth of potential for innovation and efficiency as organizations begin to embrace it. The use of these agents delivers speed and enhanced decision accuracy while empowering businesses to handle shifting environments. Through unbiased system algorithms, markets receive continuous automated predictions concerning future trends with anomaly detection to generate real-time automated recommendation solutions. To function at their peak, medical services, the financial industry, and manufacturing all require quick, decisive calculations.

Future operational possibilities are created by predictive decision-making software agents linked to business intelligence systems driven by cloud platforms. AI algorithms, scalable cloud architecture, and real-time data analytics enable enterprises to achieve exceptional operational intelligence outcomes. For organizations to reach their full potential, issues with ethics, openness, and trust must be resolved. Companies use responsible governance and frameworks for technology growth to create autonomous agents that are useful strategic transformation agents.

Multiple Choice Question (MCQs)

1. Which of the following best describes AI agents for reasoning and planning?

    A. AI agents that are designed for customer service automation

    B. AI agents that can make decisions based on logical analysis and future projections

    C. AI agents that focus on visual recognition

    D. AI agents that are used primarily for speech recognition

2. What is the primary goal of optimizing decision-making in complex scenarios using AI?

    A. To minimize the number of agents required for decision-making

    B. To make faster decisions with less accuracy

    C. To improve the quality of decisions in uncertain or complex environments

    D. To reduce the computational cost of decision-making processes

3. Which of the following is a common application of autonomous decision-making agents in strategic business decisions?

    A. Optimizing product prices in real-time

    B. Performing manual inventory management

    C. Generating social media content

    D. Providing customer feedback

4. How do AI agents contribute to healthcare decision-making?

    A. By diagnosing diseases based solely on patient history

    B. By replacing human doctors in all decision-making processes

    C. By assisting doctors with data analysis and recommending treatment options

    D. By automating all administrative tasks in hospitals

5. What is one key challenge in optimizing decision-making in complex scenarios with AI?

    A. The AI's inability to process large amounts of data

    B. The difficulty in defining the decision-making process for AI agents

    C. Lack of available data for decision-making

    D. AI's reliance on simple decision trees

6. Which type of AI technology is often used for reasoning and planning in decision-making agents?

    A. Neural networks

    B. Rule-based systems

    C. Evolutionary algorithms

    D. Genetic algorithms

7. In the context of strategic business decisions, what role does AI play in forecasting market trends?

    A. AI predicts future trends with complete certainty

    B. AI provides real-time data to assist decision-makers in understanding market dynamics

    C. AI makes decisions without human intervention in all cases

    D. AI focuses solely on customer satisfaction

8. **What is the role of AI in enhancing patient outcomes in healthcare decision-making?**

    A. AI replaces doctors for all patient diagnoses
    B. AI helps doctors by analyzing large datasets and suggesting personalized treatment plans
    C. AI eliminates the need for patient consent in medical procedures
    D. AI handles administrative tasks like billing and insurance claims

9. **Which factor is most critical for the successful deployment of autonomous decision-making agents in complex scenarios?**

    A. The agent's ability to process data quickly
    B. The agent's ability to make decisions with no human input
    C. The agent's ability to adapt to new data and changing conditions
    D. The agent's reliance on historical data only

10. **In healthcare, AI agents are most useful when they:**

    A. Perform surgeries autonomously
    B. Automate all decision-making without human oversight
    C. Support healthcare professionals by providing insights based on patient data
    D. Replace healthcare professionals in all areas of patient care

Answer

1	2	3	4	5	6	7	8	9	10
B	B	A	B	C	B	B	B	C	C

PART IV

Advanced Topics in Generative AI Agents

CHAPTER 13

Scaling Generative AI Agents

13.1 Infrastructure Considerations for Scaling AI

The Generative AI's quick developments have generated a lot of curiosity and enthusiasm since they provide previously unheard-of opportunities for creativity, productivity, and problem-solving. To fully realize generative AI's capabilities, one must build infrastructure capable of sustaining complex model training operations along with their execution requirements.

Figure 13.1: This modern server demonstrates blue day-glow lighting with circular elements and related server shapes distributed across its framework. Geometric patterns generate projections from these servers.

Source: – (Clouds 2023)

Companies use Cloud computing as their main platform to develop AI systems which lets them access flexible, scalable resources at cost-effective rates. Organizations gain access to strong computational power through cloud systems without spending large capital amounts at the beginning (Clouds 2023).

Benefits of Cloud Computing for Generative AI

The development of generative AI applications receives multiple advantages from cloud computing deployments:

- **Scalability:** Cloud platforms show excellent adaptability to varying computational demands of generative AI models while providing effective resource distribution methods. The training requirements of models call for individual focus when one works with cloud-based platforms.
- **Cost-effectiveness:** Organizations reduce costs by using cloud computing›s payment system for usage instead of needing to buy expensive hardware. Entities that need flexible working requirements or financial constraints see the greatest advantages from pay-per-use cloud computing systems.
- **Flexibility:** Generative AI›s quick developments have generated a lot of curiosity and enthusiasm since they provide previously unheard-of opportunities for creativity, productivity, and problem-solving.
- **Time-to-market:** World-class resource management solutions combined with rapid model deployment drill down the production schedule when developing generative AI solutions. The ability to rapidly provision resources while deploying models plays a crucial role in current competitive markets that prioritize rapid delivery.

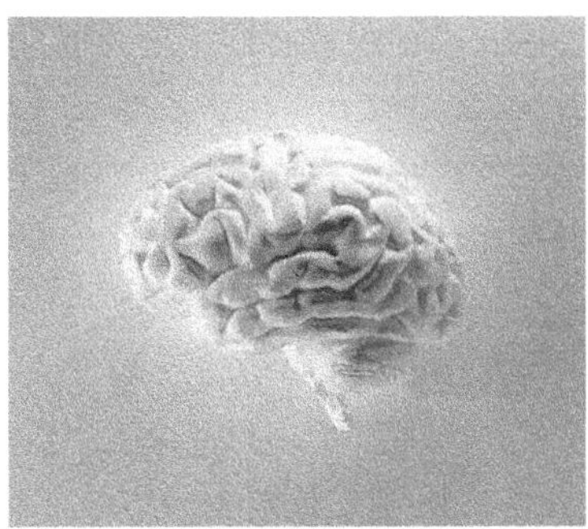

Figure 13.2: A three-dimensional human brain model receives illumination from a soft purple-blue light to create its modernized knowledge-oriented appearance in this visual representation.

Source: – (Clouds 2023)

Essential Components for Generative AI in the Cloud

Establishing a secure generative AI solution for the cloud requires picking specific components able to meet the processing needs, along with the data requirements of complex models. (LeewayHertz 2023).

The following essential points demand attention:

a. **Compute Resources:**

- **Central Processing Units (CPUs):** Deep neural network training demands special hardware optimization due to its intensive parallel operations, although CPUs perform well across multiple computing needs. Certain generative AI pipelines benefit from the versatile use of CPUs, although they lack optimal efficiency in parallel tasks.
- **Graphics Processing Units (GPUs):** The processing program of GPUs functions efficiently with parallel systems through specialized design features specifically for matrix operations and image processing. Generation AI workloads benefit from GPUs which have emerged as a preferred technology option.

- **Tensor Processing Units (TPUs):** The specialized hardware accelerator TPUs offers exceptional machine learning performance to execute neural network training together with inference operations. The combination of performance improvement alongside energy savings through accelerators makes AI workloads suitable for scalable generative AI deployments.

b. **Storage:**

- **Object Storage:** Big datasets and trained models can be managed efficiently through the unstructured data management provided by the storage platform, which boosts storage possibilities.
- **Block Storage:** Users can locate persistent storage for their files along with directory structures through Block storage to meet their requirements. Model training sessions allow users to save intermediary information and checkpoint data through Block storage.

c. **Networking:**

- **High-speed network connectivity:** A reliable high-speed network maintains smooth data transfer needed to support communicative operations inside generative artificial intelligence systems.
- **Network optimization:** The combination of load balancing systems and content delivery networks enables network performance improvements and latency reduction support.

d. **Data Management Tools:**

- **Data lakes:** Modern organizations establish central databases to store vast amounts of unprocessed data, which enables flexible, scalable management of diverse datasets.

- **Data pipelines:** Workflows automated to move, transform, and analyze data help maintain efficient data movement while ensuring quality control.

e. **Additional Considerations:**

  - **Scalability:** System designers must implement infrastructure capabilities that extend processing capacity because generative AI models will grow with increased complexity.
  - **Performance optimization:** Model speed and training performance optimization require applications of hardware acceleration together with model optimization and hyperparameter tuning.

13.2   Distributed Systems for Large-Scale Agents

Due to the quickly progressing generative AI, the problem of complexity and scale of agents requires a more efficient distributed system. Since these agents are going to be applied and inserted into various fields such as customer service, producing creative content, and autonomous entities, the core framework of this architecture should be flexible, dependable, and optimal.(Fabrice Bouquet 2015). In this chapter, we discuss the fundamental elements and approaches to the architecture of DSs for massive-scale generative AI agents.

  ➢ **Architectural Considerations**

  Subsequently, modularity can be considered an essential principle defining distributed systems for generative AI. The use of a microservices architecture makes it possible for the developers to decompose the many functionalities of AI agents into smaller services. For example, data pre-processing, model training and inference, and feedback loop processing can be done by discrete services. In addition to the improvement in scalability, this does the same for maintainability as the architecture can be updated and changed easily, especially since it is divided into modules.(Feng et al. 2024)the problem of deployment latency optimization can be transformed into an NP-hard mathematical optimization problem. However, in the real world, deployment strategies at the edge often require immediate updates, while human-engineered code tends to be lagging. To bridge this gap, we innovatively integrated LLMs into the decision-making process for microservice deployment. Initially, we constructed a private Retrieval Augmented Generation (RAG.

Another promising model is serverless computing, which is rather useful for an unpredictable or occasional processing model. It can also be taken and articulated that the solution is cost-effective since organizations will not require investing in actual servers to host the applications. Also, Event-Driven Designs principles, for instance, the use of message queues or event streams, help to have the flexibility in the sequence of 'practical' actions that were designed to increase the 'responsiveness' of the system.

➢ **Achieving Scalability**

Scalability is one of the fundamental requirements in any distributed system to be employed for large-scale generative AI. Horizontal scaling, that is the process of adding new nodes in order to accommodate more workloads is a basic technique. Traffic distributing devices are very important in serving purposes with other servers, avoiding the occurrence of a particular server being loaded beyond its resource capacity.

In the large actualization models, model sharding can be used to distribute computational loads across a server or GPU. This approach not only enhances performance, but also enables big, complex models that are out of reach of a single machine to be deployed,(Cui et al. 2016)training such networks on a single GPU is too slow and training on distributed GPUs can be inefficient, due to data movement overheads, GPU stalls, and limited GPU memory. This paper describes a new parameter server, called GeePS, that supports scalable deep learning across GPUs distributed among multiple machines, overcoming these obstacles. We show that GeePS enables a state-of-the-art single-node GPU implementation to scale well, such as to 13 times the number of training images processed per second on 16 machines (relative to the original optimized single-node code. Other features, such as dynamic scaling and autoscaling, go a step further to increase the flow balance and efficiency of resource usage without overconsuming resources.

> **Data Management Strategies**

Data management is a critical aspect that determines the reliability of the generative AI systems in use today. Structured data and records are efficiently processed by distributed databases, including Apache Cassandra or MongoDB, since the mentioned systems entail the management of huge amounts of data. These databases provide high availability and fault tolerance, advantageous characteristics in large implementations.

Some of the cache techniques, like Redis / Memcached, can contribute greatly in efficiency of data fetch especially data that is often used. Extenz Data partitioning effectively breaks down data according to parameters such as geography or user type, which in turn maximizes performance by limiting latency while also restricting overall demand on servers,(Custer 2023). Cumulatively, the strategies guarantee that the system can accommodate the real-time data feed processing.

➢ Model Management and Optimization

The use of distributed system to manage the bigger AI models has some other complexities. Training model, also called model parallelism where a model is cut into halves and processed on two GPUs or TPUs at the same time shortens the inference time. In the decentralized cases, federated learning enables the training of the models near the source of data, which decreases the demand for the centralization of data and increases privacy (Kumar 2023).

Default tools for versioning are MLflow or DVC as they help in tracking multiple iterations of a model. It also extends the capability of those tools to allow teams to test and launch new features with minimal disruption to productivity. This capability is critical in dealing with generative AI agents which are periodically tuned to the need and the environment they are in.

➢ Communication Between Agents

Coordination of distributed AI agents requires an effective mode of communication. For example, Apache Kafka or RabbitMQ allow achieving reliable inter-agent communication using so-called message queues. It also implies that these systems facilitate asynchronous communications, which are critical to the throughput and inter-message response times of large-scale deployments.

Interfaces, especially gRPC/REST, help in communication between microservices and agents, and since gRPC supports streaming call types such as bi-Directional streaming, it is well suitable for many kinds of real-time communication. These tools allow easy interconnection of the parts of the system to facilitate the overall functioning of the distributed system as a single system.

➢ **Security and Privacy**

One of the biggest challenges for generative AI in distributed systems is security. A proper choice of security measures like OAuth 2.0 in the context of authentication and authorization provides better access control with system resources. The protection of data in motion and data at rest preserves confidentiality as earlier pointed out.

There are applications that are designed to work as anomaly detection systems that can alert for such things as break in attempts or suspicious use of resources within the network. It is some additional levels of protection, which allow the alerts about potential threats to be seen and prevented.

13.3   Managing Latency and Real-Time Performance

There are several applications for real-time generative AI, which produces material instantly. It demonstrates its versatility by powering chatbots for customer service and assisting in the creation of original content. To get the most out of real-time generative AI systems, we must understand what it can and cannot achieve. This well-rounded perspective enables us to create innovative and captivating applications for it.

Figure 13.3: A working professional leveraging on futuristic Smart Technologies in Decision making supported by Real-Time Data Analysis.

Source: – (LABS 2024)

Challenges of Real-Time Generative AI

> **Latency and Response Time:**

- Real-time applications require fast reactions. When a generative AI program has to do intricate calculations, it might produce content, which can slow down operations and make real-time usage challenging.
- Speeding up replies can be achieved by reducing the size of models, eliminating extraneous components, and utilizing specialized technology. According to a research, inference time might be lowered by 40–60% by optimizing a large-scale generative AI model for TPUs. (Silva and Costa 2023)virtual assistants, and machine learning, among others. This evolution is taking place in various areas such as health, finance, marketing, and certainly in the field of human resource management.

The goal is to help automate routine tasks, improve the efficiency and effectiveness of processes, and provide valuable outputs for strategic decision-making.\nThe advent of ChatGPT has further strengthened support for the Human Resources area, having a very wide-ranging application, but which can also be personalized according to the real needs of the organization. But what is ChatGPT? It is a language model created by Open AI and based on GPT – Generative Pretrained Transformer, (an unsupervised language neural network that was trained using massive amounts of internet data.

➢ **Computational Resources**

- **Resource-hungry models:** In order to learn and function, generative AI systems that create fresh, meaningful content require a lot of processing power.
- **More hardware:** The scale and complexity of real-time AI applications may be constrained by the number of machines (CPUs, GPUs, and TPUs) that are available.
- **Using the cloud:** Through cloud platform connections, users gain access to additional computing power when required. OpenAI reports that training a large-scale generative AI model requires thousands of GPUs.

➢ **Data Limitations**

- **Data quality and quantity:** Enhanced generative AI model performance emerges when training data reaches both respectable quality and size benchmarks.
- **Data privacy:** Public concern about privacy emerges when large data collections undergo utilization.
- **Data augmentation:** Models operate reliably across diverse situations by using augmentation methods even beyond their starting data limitations. Stanford

University researchers demonstrated that data augmentation increased image classification model accuracy by between 5 and 10 percent through their examination.

➢ **Ethical Considerations**

- **Bias and fairness:** Foundation models create outputs that contain unlawful bias and discriminatory content because they rely on their training data.
- **Misinformation and deep fake:** False content has eroded people's trust because Generative AI produces fake material that disarms people into accepting the simulation as true.
- **Transparency and explainability:** Research shows that users show a 77% level of concern about bias within AI system datasets. No artificial intelligence modeling choice implementation can be effective without testing system responsibility while maintaining fair system choices. According to Pew Research Center studies 77% of people express anxiety about AI systems possibly acting with bias.

Techniques for Optimizing Real-Time Performance

1. Model Optimization:

The mathematical computations at Future Estystems become more efficient through their adoption of model simplification methodology.

- **Quantization:** The model needs reduced numerical precision for faster computation times and easier system resource calculations.
- **Distillation:** Information transfer from huge, complicated models to smaller, more effective ones.

2. Hardware Acceleration:

- **GPUs:** Computational matrix operations alongside standard deep learning calculations execute at accelerated speeds through parallel processing performed by Graphics Processing Units.
- **TPUs:** Machine learning hardware devices called Tensor Processing Units enhance the efficiency of specific tasks during computer operations.

3. Cloud-Based Infrastructure:

- **Scalability:** Cloud-based systems rapidly extend their resource capacity to meet requirements for real-time applications.
- **Cost-efficiency:** Businesses can handle shifting workflow costs effectively through their pay-as-you-go pricing structure.

4. Efficient Data Pipelines:

- **Batch processing:** This data processing method provides faster operational speed for handling data batches.
- **Streaming processing:** This system performs data processing in real-time operation.
- **Data caching:** The technique speeds up data retrieval by often storing data points in memory.

Industry-Specific Applications

- **E-commerce:** Product descriptions, item recommendations, and customized marketing strategies.
- **Finance:** Creating artificial financial data to evaluate risk and train fraud detection programs.
- **Education:** Creating individualized learning resources and assessments.
- **Manufacturing:** Improving product design and optimizing production procedures.

Businesses from a variety of industries can discover new methods to expand, increase production, and satisfy consumers when they make use of generative AI's potential.

Figure 13.4: The above image shows a man wearing a black suit and tie standing beside a laptop computer. He is discussing something with two hands as if he is in front of someone. Around him, there are several logos and icons related to artificial intelligence, which mark that he is working with it.

Source: – (LABS 2024)

13.4 Cost and Resource Optimization

The problem of scalability is especially crucial when it comes to actual generative AI agents: it is difficult to achieve both cost-efficient and resource-efficient, high-performing, and reliable models. Unsupervised generative AI, like the large language models or image synthesis models, are computationally heavy models that might need large memory as well as large training data resources. Indeed, the scaling strategies highlighted earlier have a common imperative; they need to be technological as well as economic.

One other area of significance for improving the use of resources would be model and model compression and fine-tuning. Reducing the footprint and computational intensity of generative structures, methods such as quantization and pruning, as well as knowledge distillation.

Quantization means a process in which all the weights and activations of the model are retained with less precision, thus using minimal space and performing computations faster. The process of converting the information of a large, complicated model to a smaller model that performs similarly but requires fewer calculations is known as knowledge distillation. Another way of saving costs and limiting resource usage is when training models for particular tests, instead of starting from scratch, is to use pre-training models. These techniques allow organizations to establish AI agents on a larger number of devices such as edge devices so cutting down the operational costs and ease of use (Novac et al. 2021).

The last but equally important element is model compression and fine-tuning as applied to the management of resources. For example, the generative AI models size can be reduced with a quantization technique, the number of computations necessary in generative AI models can be reduced with the help of or pruning technique, and the knowledge of the generative model can be transferred with the help of a knowledge distillation technique. Quantization concept is about encoding the model weights and activations together with fewer bites as this helps in onboard memory as well as boosts computations. They are probably the most effective noise reduction process in the node; pruning eliminates extra parameters that typically clutter a given model and ultimately improve its performance. Knowledge distillation is a significantly faster way to move knowledge from a large, complicated model to a much smaller one while maintaining performance. Training costs and resource utilization are also cut down could by using, instead of training models from new, fine-tuning for targeted tasks on pre-trained models. These methods allow organizations to run NC's AI agents on substantially more devices, from edge devices, while at the same time, reducing operational costs and increasing possibility.

Another is achieving data efficiency – or optimizing data usage in and of itself. Thus, its usage offers value by presenting high-quality and well-curated datasets that may minimize the proportion of training iterations that are unnecessary and resource-consumptive. Some wish to improve the generalization capability of models without using more samples and data augmentation methods can do this by increasing the diversity of samples in the set. Another critical approach is active learning, where, for example, the model selects the data points that need labeling the most, and transfer learning, where data knowledge from one domain is transferred to another, adds even more value to data. These approaches, apart from simply helping to speed up the scaling process, also decrease the number of training runs needed on large infrastructures, thus making AI deployment a more environmentally friendly process. (Hu, Zheng, and Yang 2011).

Effective and major system monitoring coupled with adaptive resource management are critical to sustaining cost and resource efficiency at the time of scaling. The ability to track the performance of a system and resources with cost data in real-time results in a better understanding of when a change is needed. Autoscaling mechanisms, for example, are mechanisms that can scale up or scale down a resource's usage depending on traffic or computational load; that is, one can find that a resource is underutilized yet busy providing a service, or that a resource is overwhelmed yet there is a scarcity of users.

Specifically, scaling can optimize its strengths by integrating some ideas such as resource sharing and collaborations. For organizations to be involved in open-source community or federated learning frameworks they get to share the costs of compute-intensive tasks such as model training, though data privacy and security is safeguarded. Federated learning for instance allows several parties train models on decentralized data without having to centralized the data and, therefore, decrease the costs involved. This collective approach minimizes the costs on individual companies and increases innovation through access to different knowledge and experience. Pre-built tools and frameworks are freely available in open-source platforms, which not only saves time to write code for a particular module, but also prevent the development of tools that may already exist,(Yuliya Melnik 2023).

Lastly, there is room to consider the possibilities of using different types of hardware to support cost savings. GPGPUs, TPUs and otherwise Custom AI ASIC accelerated architectures such as are tailor made to execute tasks of the kinds demanded by AI applications with superior efficiency as compared to that of traditional CPUs. Actually, investing in these kinds of hardware solutions can bring lots of performance improvement as well as much lower costs in the future. Likewise, edge computing can delegate some of the computational loads to the local side, thus relieve the center side from congestion, high latency, and high bandwidth charges while enhancing real-time responsiveness.

5 Ways Generative AI Optimizes Costs with Smarter Resource Allocation

1. **Transforming Project Management:** Project management gets transformed by artificial intelligence systems that enable teams to monitor workflow progress concurrently

with resource reallocation capabilities. AI technology helps organizations detect workflow blockages so they can relocate trained employees into core project areas. AI Project Management shows remarkable business growth as analysts expect this trend to continue, achieving a 15 percent+ Compound Annual Growth Rate from 2024 to 2032. Various business functions require more efficient ways to manage projects and predictive analytics capabilities, thus driving this market growth.

Through its data processing capabilities, GenAI outperforms human analytical capability by producing quick, efficient results for comprehensive analysis. The system integrates advanced processing components by uniting team capacity analysis with project requirement diligence alongside strict deadline restrictions and historical project archival data to deliver sophisticated data-based outcomes. The comprehensive data approach enabled by this methodology leads to decisions made with enhanced accuracy and speed, as well as traditional deficit reduction.

2. **Driving Cost Efficiency:** The software industry achieves its highest financial productivity gains along with improved productivity within domains enabled by GenAI technology. GenAI technology generates exceptional content products that accelerate content development methods. The adoption of AI tools throughout development activities has resulted in developers doubling their coding efficiency, and surveys from McKinsey reveal program development speeds increase by 100%. Operations benefit from automated tasks and predictive abilities, and optimized resource distribution which results in significant cost reductions. The product design process experiences transformation thanks to GenAI platforms which 32 percent of companies utilize to develop prototypes while

generating new concepts. These platforms let users explore brief solutions that can develop precise specifications through complete design refinement. AI integration with modern development tools creates agile operational methods which drive innovation and modify software enterprise operations alongside relaying market competition.

3. **Enhancing Decision-Making and Greater Efficiency:** GenAI innovation revolutionizes organization-based decision-making processes. Business solutions based on advanced artificial intelligence leverage data recommendations to enhance strategic choices for pricing decisions and competitive data reviews as well as cost assessments and product strategy development. Strategic integration of GenAI technology provides organizations with financial cost reductions exceeding 30 percent. Responsible implementation of this technology establishes a valuable business asset that improves innovation and creates industry competitiveness.

4. **Customer Service Revolution:** Logically driven technology solutions like virtual assistants resolve simple customer inquiries fast which leaves support staff to manage more intricate issues. Organizations using GenAI for their pricing strategy can analyze broad market data and competitor activities along with customer patterns to drive improved revenue gains. According to McKinsey's digital budget surveys AI technologies represent a minimum of 20% of total digital investment while McKinsey research identifies pricing strategies as fundamental priorities. Organizations using data-led teams utilize advanced algorithm implementations to build pricing models that boost marketing performance and increase sales outcomes.

5. **Code Enhancement:** Research indicates that the most remarkable aspect of Generative AI emerges into view as its

primary function. GenAI stands among its fellow AI systems because it provides additional functionality by improving existing source code. AI models identify programming patterns to create advance defect detection systems that stop severe system disruptions from occurring. Digital predictive models which use predictive maintenance enable developers to envision upcoming software upkeep requirements while simultaneously detecting outdated libraries and code compliance violations. Organizations can maintain operational system availability before operational disruptions cost them money. Leading cybersecurity organizations now operate under a built-in proactive security model as their standard operational practice. Real-time security vulnerability identification occurs through AI support for code and system log assessments which advances sponsored prevention methods. The deployment of artificial intelligence capabilities extends past security functions within these applications. These structural enhancements come from operational interactions between intelligent algorithms and security assessment activities.

To expand AI generation systems companies, need robust infrastructure and responsible practices and proper organizational alignment. Advanced generative AI agents in combination with scenario simulation and massive problem-solving capabilities enable true content creation and drive operational transformations across business fields involving marketing and product design with customer service applications. Many organizations encounter substantial hurdles in their effort to scale this technology because of operational challenges in implementing computational resources optimization alongside model drift control solutions and data privacy compliance requirements. Organizations reach successful generative AI agent scaling through the combination of cloud platforms with distributed computing and adaptive AI learning methods that improve operational accuracy and efficiency and accountability. Strategic application of these agents generates well-defined practical methods that empower organizations to create innovative industrial solutions while establishing future market leadership positions in the modern digital age.

Multiple Choice Question (MCQs)

1. What is the primary concern when considering infrastructure for scaling generative AI agents?

    A. Network speed
    B. Hardware and software resources
    C. User interface design
    D. Agent creativity

2. Which of the following is a key component of distributed systems for large-scale AI agents?

    A. Centralized data storage
    B. Real-time data processing
    C. Single-point data entry
    D. Autonomous learning

3. What does "latency" refer to in the context of generative AI agents?

    A. The amount of data processed by the agent
    B. The delay between the input and the response from the agent
    C. The memory capacity of the agent
    D. The training time of the AI model

4. Which of the following strategies helps reduce latency in large-scale AI systems?

    A. Using complex neural networks
    B. Increasing the number of data layers
    C. Employing edge computing for local processing
    D. Reducing the frequency of updates

5. **When scaling AI agents, which factor primarily affects the real-time performance of the system?**

    A. Number of training epochs
    B. Data synchronization and transfer speed
    C. User interface complexity
    D. AI model transparency

6. **Which of the following is an example of cost optimization in AI systems?**

    A. Increasing the number of training parameters
    B. Minimizing the use of cloud-based services
    C. Leveraging batch processing for large datasets
    D. Utilizing more powerful hardware without efficiency considerations

7. **What is the role of distributed systems in scaling generative AI agents?**

    A. They help maintain a single source of data
    B. They enable parallel processing of data across multiple machines
    C. They reduce the need for training models
    D. They focus on user interaction design

8. **What is one of the challenges of scaling generative AI agents in terms of resource optimization?**

    A. The complexity of training the model
    B. The need for constant model updates
    C. Managing energy consumption and computational costs
    D. Reducing the size of the training dataset

9. In a distributed AI system, which of the following is critical for ensuring smooth communication between nodes?

A. Uniform memory access
B. Low-latency network protocols
C. High-definition graphics processing
D. Centralized control of all data

10. What does real-time performance in AI agents mainly depend on?

A. The quality of the training dataset
B. The number of parameters in the model
C. The speed of data processing and response time
D. The complexity of the AI's decision-making process

Answer

1	2	3	4	5	6	7	8	9	10
B	B	C	B	B	B	C	C	B	C

CHAPTER 14

The Future of Generative AI Agents

14.1 Towards Artificial General Intelligence (AGI)

The field of Artificial General Intelligence (AGI) brings together artificial interactivity models which attain human-level mental competence across a diverse set of capabilities. Agents with Artificial General Intelligence (AGI) aspire to show adaptable intellectual capabilities since their algorithms focus on solving different issues beyond set instructions in the fashion of narrow AI. (Li 2023)I argue that given the possibility and prospect of Artificial General Intelligence (AGI.

Comparison to Narrow AI (ANI)

A proper grasp of AGI requires researchers to distinguish it from the category of narrow AI (ANI). Artificial Narrow Intelligence otherwise known as Narrow AI is built to perform specific tasks. Examples of narrow AI include:

- ➢ **Speech Recognition:** Programs such as Siri and Google Assistant serve as models that detect spoken commands and convert them into usable information.
- ➢ **Image Recognition:** The technology enables medical imaging solutions alongside facial recognition systems that analyze visual patterns within data.
- ➢ **Recommendation Systems:** Recently developed algorithms help Netflix and Amazon recommend content and products through monitored user preferences.

The fundamental differences between AGI and ANI are:

- ➢ **Scope of Functionality:** ANI functions through specific assignments but AGI incorporates capabilities to perform various complex assignments matching those normally achieved by human intelligence.
- ➢ **Adaptability:** Specific task programming stands as a necessity for ANI yet AGI achieves task adaptability and environment autonomy without requiring fundamental programming.
- ➢ **Cognitive Abilities:** The goal of AGI technology is to duplicate aspects of human cognitive ability that describe learning processes along with reasoning methods and decision-making capabilities superior to what ANI can achieve.

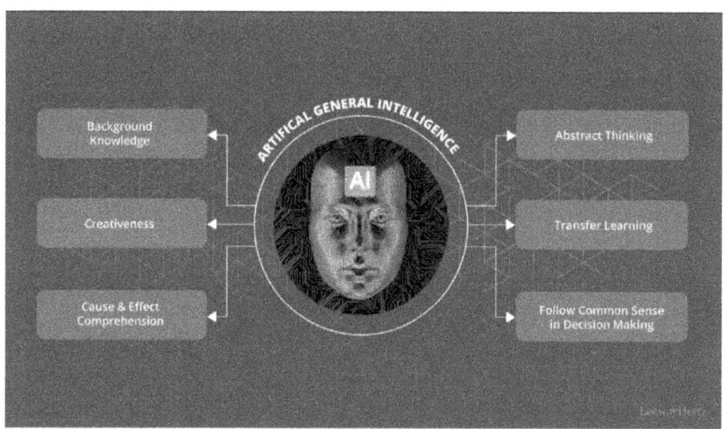

Figure 14.1: The above image represents fundamental AGI functions: These functions include background knowledge along with creativity and causality understanding, as well as abilities for abstract reasoning and transfer learning, and practical judgment.

Source: – (Akash Takyar 2023)

Unprecedented progress and difficulties emerge because of AGI's full range capability to accomplish intellectual tasks that humans can do. Expanding beyond technological machine advancement represents a fundamental breakthrough within our technological exchanges. We need to research technical aspects alongside the ethical implications and social ramifications of AGI development, which benefits all humans. (PurpleSlate 2024).

Key Characteristics of AGI

- **Versatility**

  One of the defining features of Artificial General Intelligence (AGI) is its versatility. The broad operation characteristics of Artificial General Intelligence systems differ from narrow AI systems through its ability to execute diverse activities. The adaptable capabilities of AGI technology permit it to move between tasks with different levels of difficulty without needing extra programming.

  The system with artificial general intelligence operates across administrative tasks and creative projects in addition to difficult analytical operations under a single operational system and platform. This flexible platform shows strong functionality across different applications which enhances professional utility in their independent work areas.

- **Learning and Adaptation**

  These systems apply a design strategy to process experience data for future operational development. Through its adaptive nature AGI continually extends performance capabilities during execution phases to produce successful outcomes from newly detected operational events. The autonomous acquisition of new information enables AGI to grow its operational competence and method capabilities making it more capable.

  Design paradigm derives modified system behavior through the analysis of accumulated system history data. Operating agents with adaptive systems build better performance skills when completing tasks, they originally knew nothing about. Systematic learning through AGI results in improved procedural methods while enhancing its task understanding capabilities which improves system functionality.

- **Reasoning and Problem-Solving**

An AGI system's special capability enables unknown solution processing through its capacity to integrate logical assessment with problem-solving properties. The fundamental workings of overall problem resolution move past fixed procedure types to let AGI technology generate responses which extend beyond coded instructions. Through its accessible features, AGI systems resolve hard challenges without needing human supervision at any point.

The AGI system demonstrates its capability through big data processing to discover hidden patterns that lead to innovative useful solutions. Autonomous AGI systems need problems solving capabilities to handle independent tasks across numerous subject fields.

- **Autonomy**

Autonomous operation functions on the foundation of AGI as a platform which needs no ongoing human operations for support. The learnable autonomous adaptation abilities built into AGI systems lead to independent decision making across various encounters. For AGI to perform tasks effectively and efficiently in real-world operational settings it needs autonomy.

Under autonomous operation AGI systems control entire operational workflows throughout the analytical cycles of data collection and analysis and decision-point execution.

Through autonomous operation AGI significantly improves its operational efficiency while decreasing human intervention so users can dedicate themselves to strategic and creative pursuits.

Theoretical Approaches to AGI

- **Symbolic AI**

  Symbolic Since artificial intelligence uses symbols to mimic human intellect, it functions using classical techniques. The approach relies on knowledge representations and explicit programming principles. Styled AI systems may solve issues and reason using natural language processing because they use logic-based frameworks for information representation and processing.

  The first type of artificial intelligence examples consists of expert systems alongside early programs IBM Watson employed a vast database for competing on Jeopardy.

- **Connectionist Approaches (Neural Networks)**

  Neural networks are used in connectionist approaches to artificial intelligence, and its notion is based on the unique structure and functions of human brains. According to this method, neurons in each processing system work in parallel to exchange data at linked nodes that are utilized across several layers. Because neural networks can learn from data, they are very useful for tasks like audio processing, picture identification, and natural language comprehension.

  Examples of connectionist AI include deep neural networks used in autonomous driving systems and deep learning models used in applications such as Google's AlphaGo, which defeated a human world champion in the game of Go.

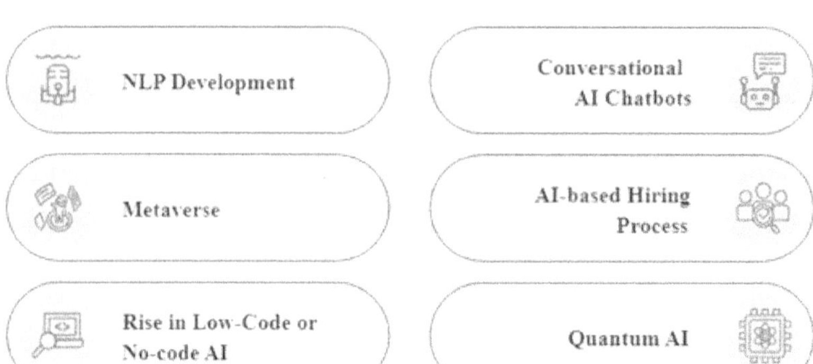

Figure 14.2: Latest trends in AGI.

Source: – (Akash Takyar 2023)

<u>The Role of AI in Shaping the Future of Work</u>

The generative AI agents are best suited because they perform routine and time-consuming duties to relieve employees of uninteresting assignments. In areas like marketing, customer support or software engineering they can create textual material, attend to elementary questions, write programs. While automation threat hooding of genuine employment, it also sets onset for new positions centered purely on idea creation, critical analysis, and planning (Akash Takyar 2022).

14.2 Case Study: AI in Content Creation

The case of generative AI that is used, for instance, for article writing and graphic design demonstrates how creativity can be extended through use of such tools. Thus, having their essentials covered by other solutions, professionals can focus on ideas and how they are presented to the public. For instance, in the marketing sector, AI can be applied in writing promotional messages which the marketing team will probably modify while ensuring they meet the brand's ethos.

- **Growing Dynamics and Efficiency**

  AI is revolutionizing interaction because agents act as assistants and facilitators of communication in group settings. Such systems can call for a meeting, take notes during the discussion and generate decisions based on the analyzed data. The quick processing of Big Data appeals to Key Decision-makers as it allows teams to swiftly make the right decisions which subsequently will help teams stay adaptable in volatile markets.

- **Artificial Intelligence Based Collaboration Tools**

  Members of generative AI agents within enterprise communication interfaces are automating tasks. For example, one can use the tools powered by AI to make summaries of many string conversations, such as emails, or automatically create the schedule of meetings. Since these tools continued to evolve, they are anticipated to facilitate better connection between cross-functional teams by minimizing barriers to knowledge sharing.

- **Reskilling and Workforce Adaptation**

  The increase in the type of generative AI will require action to develop new skills among the workers. Employers require establishing a relationship between the extent of potential AI impact and the extent of human capital development strategies, including training programs that will build technical and interpersonal skills for employment in an AI-integrated world. The role of data literacy, AI ethics, and creativity and innovation as central categories of working skills will grow.

- **The Importance of Lifelong Learning**

  Therefore, people at workplaces need to commit to continual learning, especially if they are to keep up with the competitive world. For employees to remain relevant in today's workforce, organizations have to offer online courses, certifications and other support for corporate training programs. Businesses which put efforts towards training their employees are not only safeguarding their human capital for the future Ethical Considerations and Governance.

The scalability of generative AI agents brings with it many ethical concerns. Concerns are balance, data privacy and quality, and it is always important to encourage the right use of data. Strong governance systems are necessary to guarantee that the use of AI innovation and systems is suitable and skillfully planned, which is a crucial requirement for every business or organization.

- **Building Ethical AI Systems**

Since the existences of generative AI agents can potentially infringe on people's rights, developers and organizations have to subscribe to ethical standards when creating the entities. This involve carrying out a regular audit, applying biases check system and adhering to data protection act. There is need for interconnectivity of stakeholders in the industry, the policymakers and academics to set norms that would be useful to the society.

- **The Future of Work with AI**

AI agents are not automated instruments; they are enablers for recreating work processes. These agents bring efficiency and innovation mechanisms to organizations by automating recurring tasks, encouraging creativity, and providing the big-picture decision-support. However preparing the workplace for the shift to AI needs more than a reactive approach to face the challenges that this transforming will bring as well as to effectively capitalize on the opportunities (Boomi 2024).

- **Balancing Human and Machine Contributions**

  The main issue in implementing generative AI at the workplace is to achieve an optimal combination of human and artificial intelligence. AI should supplement human skills where possible so that we receive help wherever we are weak. When combined, organizations can build the future of work that is collaborative, innovative and diverse.

14.3   Transforming Industries with AI Agents

From customer service to operational efficiency and more, AI agents are transforming business function:

Figure 14.3: The above figure indicates that AI agents are penetrating industries including financial services, healthcare, therefore, changing work and life of those industries.

Source: – (Systango 2024)

- **Finance:** Smooth operations happen through automated transaction management and suspect activity monitoring which leads to better security outcomes and increased compliance.

- **Healthcare:** Through patient data examination AI agents optimize treatment approaches and improve diagnostic techniques along with better clinical results which improves health delivery services.
- **Retail:** Artificial intelligence tools enhance industrial operations by providing customized recommendations combined with supply chain solutions to satisfy individual consumer needs.
- **Manufacturing:** Predictive maintenance analytic collaboration with AI-supplemented supply chain optimization along with quality control robotic implementation results in enhanced operational performance by minimizing equipment breakdowns.
- **Telecommunications:** The incorporation of predictive failure prevention which modeling tools offer for telecommunications systems allows network data analysis alongside optimized bandwidth distribution and customized solutions.
- **Transportation and Logistics:** Advancements that marry response route management and real time system alerts with prediction-based equipment maintenance and delivery status monitoring through the supply chain result in price efficiency in the transportation and logistics sector.

AI agents bring powerful transformation to how business operations are done through the capabilities of customer service models and operability as well as new functional applications.

1. Customer Service

Virtual service stations, together with robot agents, are used to transform the service by the agents:

Fast response to queries and complaints during extended 24-hour service period is something that makes the virtual assistants achieve superb customer satisfaction. Virtual Assistant technology will only perform at its best if User command interpretation is enabled thus improving interactive relations and shortening work timelines.

2. Sales and Marketing

Artificial intelligence agents' power changes the direction of modern sales approaches and modern marketing strategies into the form of revolutionizing operational methods:

Organizations use AI technology to evaluate the personal preferences of customers to customize marketing campaigns and assess what delivers the best rates of engagement and conversion outcomes. The analytical functionality of artificial intelligence is employed to aid in the sales execution by using its functionality to help organizations select strategic outreach best optimally and resource distribution characteristics best optimally.

3. Supply Chain Management

Artificial intelligence agents applied in supply chain management practice are several advantages:

The organization can control inventory variation, and to maintain the inventory in a way that minimizes both excess stock and product shortages, using their predictive modeling.

Using AI agents, the implementation of simplified logistical management and a decreased operational expense while speeding up operational processing activities.

4. Financial Services

Financial services organization experience expanded security measures while maintaining operational efficiency:

Through their ability to identify patterns of fraud they produce early notifications to stop financial loss from occurring. Through risk portfolio management AI agents maintain compliance regulations which builds trust with stakeholders at the same time.

14.4 Ethical and Social Challenges of AGI

Holding scientists accountable and transparent about AGI is vital to the concerns pertaining intelligent use of systems. This implies having methods to track decisions made by the AI software, having methods to declare sources of data and structures used in developing the AGI system, and having methods of holding developers and operators of the AGI system liable for the activities and consequences of the AGI system. Clear analytical reporting mechanisms let the stakeholders realize and investigate the actions of the AI systems, adding accountability and trust to such applications (Novac et al. 2021).

Organization of decision making is a grey area in autonomous governance of AGI systems due to the existence of a fine line between control and autonomy. The benefit of using autonomy is noted where it nurtures efficient and effective delivery of work in specified applications; autonomy, on the hand, is tinged with risks in terms of accountability, oversee by human, and other unforeseeable consequences. Robust regulation of AGI means striking the right balance of control between the human user and the machine, thus ethical principle, legal and societal values are crucial in achieving the right balance (IBM 2024).

Considering the AGI, the socio-economic impacts of the spread of which will have ethical dilemmas; job loss, and economic disparities. When robots take over people's work in different realms of the economy, people have worries over employment, income divide and access to better chances. They all demand specific activities of protectionism aimed at re-training the laid off employees, assistance to the affected regions and minimization of the adverse effects on sensitive groups. Furthermore, there should be positive policies and actions regarding the distribution of the advantages of AI for all classes, to examine assistance on the ethical utilization of AGI.

The practice of having machines and computers take on work that was hitherto performed by people can be envision to lead to massive loss of employment within the total economy. This creates an ethical question of unemployment, uncertainty and gradual elimination of jobs for tens of millions of people globally. The reduction of the effects of job displacement hence requires a foray into educational or vocational instructional upgrading for a new skill supply in the markets. More importantly, reviewing the patterned models of employment relationship, including UBI and JG, is beneficial for providing employment guarantee and developing social stability under the background of AI civilization (Steve Denning 2015).

AI technologies can only be seen to predispose a country even more towards social injustice by making levels of wealth and power to gravitate to the privileged few while at the same time making disadvantaged groups more vulnerable to injustices. Since economic disparities are being witnessed in today's world, it means that there is need to ensure that policies and interventions that are used seek to provide equal opportunities for the use of artificial intelligence, allowing the various groups to share the benefits similarly and enabling groups left behind to also benefit from the various opportunities in the digital world.

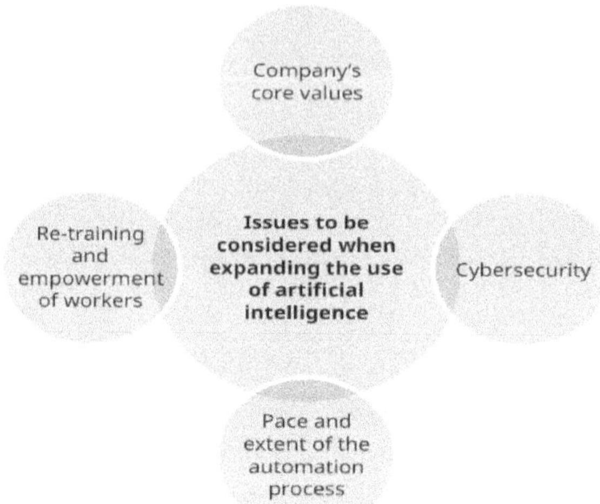

Figure 14.4: The above image shows some of the important questions that need to be asked to plan for the next steps of Artificial Intelligence use, such as the organizational values of the company, security issues, the speed and degree of automation, and how workers should be retrained and enabled.

Source: – (Able 2024)

Although the creation and research of artificial general intelligence (AGI) have great potential, they also provide several moral dilemmas that should be carefully considered. continued advancement of AGI technology requires immediate attention to ethical challenges that could produce unexpected consequences preventing beneficial uses of artificial general intelligence (Editing 2024).

- **Safety and Control:** Safety and control establishment stands as a critical ethical issue researchers face when developing highly autonomous artificial general intelligence systems. According to its definition, AGI includes autonomous operation, so scientists must develop strong control systems. Safety concerns emerge because AGI systems show the potential to produce unintended outcomes that create harm to human value systems. Researchers still need to develop protective structures in parallel with order systems that should ensure controlling dangerous autonomous AGI system behavior. According to ethical demands, maintenance of human control conditions for all AGI systems, always for the basic and complex situations.

- **Employment Disruption:** AGI systems have automatic control potential and therefore need a substantial change in the work environment. Because the highly optimized technology chooses to replace human employees, advanced generalized intelligence systems create ethical problems for human workers. There are central moral obligations on the generation of workers' need just transition and analysis of the economic impact of automation, which workers need for long-term employment security. AGI development requires parallel development of an organized assistance framework for industrial worker skill development.
- **Privacy and Surveillance:** Privacy issues arising from large-scale processing by AGI are used to legitimize surveillance. Every ethical issue is derived from determining data-driven compatibility levels relative to individual privacy rights defense. Responsible data handling and powerful data privacy protection measures are both important to the success of system operation. It's from in-depth research on Artificial General Intelligence from an integrity perspective of research that several security elements protecting data systems along with user's privacy are derived.
- **Existential Risks:** Safety problems arise in the development of autonomous artificial intelligence systems that operate without human oversight. The primary ethical approach to answering this important question is in the research protocols for AGI safety in the lab and the ethical investigations in

development. Looking ahead, research needs must be directed towards preparing control methodologies for defending against possible AGI system risks that exceed human operational capacity, since precautionary measures are directed towards limiting damaging incidents.

Multiple Choice Question (MCQs)

1. What is the ultimate goal in the development of Artificial General Intelligence (AGI)?

    A. To create AI agents capable of solving only specific tasks
    B. To develop AI that can replicate human cognitive abilities across various domains
    C. To develop AI agents for entertainment purposes
    D. To create a limited AI that focuses on a single industry

2. What is one of the major challenges in achieving AGI?

    A. Ensuring that AI agents are fast enough
    B. Ensuring that AI systems can operate with zero data
    C. Ensuring that AI systems possess generalizable intelligence like humans
    D. Ensuring AI agents have a specific domain focus

3. How is AI expected to shape the future of work?

    A. By replacing all human workers across industries
    B. By augmenting human capabilities and enhancing productivity
    C. By creating more job opportunities in manual labor
    D. By reducing the need for collaboration and teamwork

4. Which of the following is a potential benefit of AI agents in the workplace?

    A. Reduced collaboration between employees
    B. Automation of repetitive tasks, freeing up human workers for creative work
    C. Increased number of entry-level jobs
    D. Elimination of human decision-making processes

5. Which sector is least likely to be transformed by AI agents shortly?

A. Healthcare
B. Finance
C. Retail
D. Manual labor-intensive industries

6. How might AI agents transform the healthcare industry?

A. By replacing human doctors entirely
B. By providing decision support, diagnostics, and personalized treatment plans
C. By automating all medical research
D. By eliminating the need for patient consultations

7. What is one of the main ethical concerns surrounding AGI?

A. Ensuring AGI is too advanced to control
B. The environmental impact of AGI development
C. The potential for AGI to operate without human oversight
D. The financial cost of developing AGI

8. What is the social impact of AGI development likely to be?

A. Creation of a more equitable society with widespread wealth
B. Widespread unemployment due to AI taking over all jobs
C. Increased dependence on AI agents in daily life
D. Complete eradication of poverty

9. What is one potential challenge of integrating AI agents into various industries?

A. Lack of sufficient data for training AI systems
B. Lack of demand for AI solutions in most sectors
C. Resistance to change and adoption of AI technologies
D. The ability to reduce operational costs to zero

10. What role does ethics play in the development of AGI?

A. Ethics ensure AGI is developed without regard for its social implications

B. Ethics are irrelevant as long as AGI can achieve its goals

C. Ethics guide the responsible development and deployment of AGI, ensuring fairness and safety

D. Ethics only apply to the creation of AI agents for entertainment purposes

Answer

1	2	3	4	5	6	7	8	9	10
B	C	B	B	D	B	C	C	C	C

Bibliography

Able, Five. 2024. "Ethical Considerations in Artificial General Intelligence (AGI)."

Ai, Relevance. 2024. "Content Creation AI Agents." *Relevence Ai.*

Akash Takyar. 2022. "Generative AI: Use Cases, Applications, Solutions and Implementation." *LeewayHertz.*

Akash Takyar. 2023. "Artificial General Intelligence: Key Insights and Trends." *Leeway Hertz.*

Alan Zeichick. 2023. "What Is Retrieval-Augmented Generation (RAG)?" *Tech Content Strategist |.*

Alexander De Ridder. 2024. "Learn the Core Components of AI Agents." *SmythOS Academy.*

Ali, Omar, Peter A. Murray, Mujtaba Momin, Yogesh K. Dwivedi, and Tegwen Malik. 2024. "The Effects of Artificial Intelligence Applications in Educational Settings: Challenges and Strategies." *Technological Forecasting and Social Change.* doi:10.1016/j.techfore.2023.123076.

Anand Vemula. 2024. *Generative AI System Design: A Practical Guide.*

Anglen, Jesse. 2024. "Ethical AI Development Guide." *Rapid innovation.*

Anisuzzaman, D.M., Jeffrey G. Malins, Paul A. Friedman, and Zachi I. Attia. 2025. "Fine-Tuning Large Language Models for Specialized Use Cases." *Mayo Clinic Proceedings: Digital Health* 3(1): 100184. doi:10.1016/j.mcpdig.2024.11.005.

aws. 2022. "What Is OCR (Optical Character Recognition)?" *aws.*

Bandi, Ajay, Pydi Venkata Satya Ramesh Adapa, and Yudu Eswar Vinay Pratap Kumar Kuchi. 2023. "The Power of Generative

AI: A Review of Requirements, Models, Input–Output Formats, Evaluation Metrics, and Challenges." *Future Internet.* doi:10.3390/fi15080260.

Bayoudh, Khaled. 2024. "A Survey of Multimodal Hybrid Deep Learning for Computer Vision: Architectures, Applications, Trends, and Challenges." *Information Fusion.* doi:10.1016/j.inffus.2023.102217.

Belagatti, Pavan. 2024. "Multimodal RAG Using LlamaIndex, Claude 3 and SingleStore!" *Medium.*

Bhaskar, Yash. 2023. "Introduction to LLMs and the Generative AI : Part 1 — LLM Architecture, Prompt Engineering and LLM Configuration." *Medium.*

Bhutanadhu, Kusuma. 2023. "Integrating Generative AI and Reinforcement Learning for Self-Improvement."

Blog, My. 2023. "AI and Creativity Its Intersection in Art, Music, and Literature." *MY BLog.*

Boden, Margaret A. 1996. *Artificial Intelligence (Handbook of Perception and Cognition).* Academic Press.

Boomi. 2024. "How AI Is Transforming Business Process Automation."

Bozkurt, Aras. 2023. "Generative Artificial Intelligence (AI) Powered Conversational Educational Agents: The Inevitable Paradigm Shift." 18: 2023. doi:10.5281/zenodo.7716416.

Cem Dilmegani. 2024. "Top 25 Chatbot Case Studies & Success Stories [2025]."

Clifford, Ben. 2023. "Preventing AI Misuse: Current Techniques." *Centre for the governance of ai.*

Clouds, FIS. 2023. "Basic Infrastructure for Building Generative AI Solutions." *FIS clouds.*

Collins, Christopher, Denis Dennehy, Kieran Conboy, and Patrick Mikalef. 2021. "Artificial Intelligence in Information Systems

Research: A Systematic Literature Review and Research Agenda." *International Journal of Information Management*. doi:10.1016/j.ijinfomgt.2021.102383.

Cui, Henggang, Hao Zhang, Gregory R. Ganger, Phillip B. Gibbons, and Eric P. Xing. 2016. "GeePS: Scalable Deep Learning on Distributed GPUs with a GPU-Specialized Parameter Server." In *Proceedings of the 11th European Conference on Computer Systems, EuroSys 2016*, doi:10.1145/2901318.2901323.

Custer, Charlie. 2023. "What Is Data Partitioning, and How to Do It Right." *Cockroach labs*.

Díaz-Rodríguez, Natalia, Javier Del Ser, Mark Coeckelbergh, Marcos López de Prado, Enrique Herrera-Viedma, and Francisco Herrera. 2023. "Connecting the Dots in Trustworthy Artificial Intelligence: From AI Principles, Ethics, and Key Requirements to Responsible AI Systems and Regulation." *Information Fusion*. doi:10.1016/j.inffus.2023.101896.

dimension labs. 2024. "What Is a Conversational Agent."

Dimitri Didmanidze. 2024. "Natural Language Understanding (NLU) Explained."

Doanh, Doung Cong, Zdenek Dufek, Joanna Ejdys, Romualdas Ginevičius, Pawel Korzynski, Grzegorz Mazurek, Joanna Paliszkiewicz, Krzysztof Wach, and Ewa Ziemba. 2023. "Generative AI in the Manufacturing Process: Theoretical Considerations." *Engineering Management in Production and Services*. doi:10.2478/emj-2023-0029.

Dr. Jagreet Kaur Gill. 2024. "Generative AI Code Assistants for Developers."

Duan, Yanqing, John S. Edwards, and Yogesh K Dwivedi. 2019. "Artificial Intelligence for Decision Making in the Era of Big Data – Evolution, Challenges and Research Agenda." *International Journal of Information Management* 48: 63–71. doi:10.1016/j.ijinfomgt.2019.01.021.

Editing, Falcon Scientific. 2024. "Ethical Challenges in Artificial General Intelligence Research." *Falcon scientific editing*.

Eliza. 2023. "History of Chatbots: From ELIZA to Advanced AI Assistants." *Eliza*.

Emmanuel Osamuyimen Eboigbe, Oluwatoyin Ajoke Farayola, Funmilola Olatundun Olatoye, Obiageli Chinwe Nnabugwu, and Chibuike Daraojimba. 2023. "BUSINESS INTELLIGENCE TRANSFORMATION THROUGH AI AND DATA ANALYTICS." *Engineering Science & Technology Journal*. doi:10.51594/estj. v4i5.616.

Fabrice Bouquet, Patrick Taillandier. 2015. *Agent Paradigm*.

Feng, Kan, Lijun Luo, Yongjun Xia, Bin Luo, Xingfeng He, Kaihong Li, Zhiyong Zha, Bo Xu, and Kai Peng. 2024. "Optimizing Microservice Deployment in Edge Computing with Large Language Models: Integrating Retrieval Augmented Generation and Chain of Thought Techniques." *Symmetry* 16(11). doi:10.3390/sym16111470.

Ferrer, Josep. 2024. "Fine-Tuning LLMs: A Guide With Examples."

Frank van Harmelen, Vladimir Lifschitz, Bruce Porter. 2008. *Handbook of Knowledge Representation (Foundations of Artificial Intelligence) (Volume 1)*.

Geeksfgeeks. 2024. "What Is a Neural Network?" *geeksfgeeks*.

geeksofgeeks. 2024. "AI Ethics : Challenges, Importance, and Future." *geeksofgeeks*.

Gill, Dr. Jagreet Kaur. 2024. "Differences Between NLP vs. NLU vs. NLG."

Górriz, J. M., I. Álvarez-Illán, A. Álvarez-Marquina, J. E. Arco, M. Atzmueller, F. Ballarini, E. Barakova, et al. 2023. "Computational Approaches to Explainable Artificial Intelligence: Advances in Theory, Applications and Trends." *Information Fusion*. doi:10.1016/j.inffus.2023.101945.

Gumińska, Urszula, Aneta Poniszewska-Maranda, and Joanna Ochelska-Mierzejewska. 2023. "Trends in Natural Language Understanding with Sentence Representation and Sentiment Analysis." doi:10.21203/rs.3.rs-3536120/v1.

Gupta, Priyanka, Bosheng Ding, Chong Guan, and Ding Ding. 2024. "Generative AI: A Systematic Review Using Topic Modelling Techniques." *Data and Information Management*. doi:10.1016/j.dim.2024.100066.

Harshvardhan, GM, Mahendra Kumar Gourisaria, Manjusha Pandey, and Siddharth Swarup Rautaray. 2020. "A Comprehensive Survey and Analysis of Generative Models in Machine Learning." *Computer Science Review*. doi:10.1016/j.cosrev.2020.100285.

Hu, Derek Hao, Vincent Wenchen Zheng, and Qiang Yang. 2011. "Cross-Domain Activity Recognition via Transfer Learning." *Pervasive and Mobile Computing*. doi:10.1016/j.pmcj.2010.11.005.

IBM. 2024. "What Is Artificial Intelligence (AI)?" *IBM*.

Idan Novogroder. 2024. "Top 8 AI Frameworks: Benefits & How to Choose the Right One."

Jaidev, Uma Pricilda, and Susan Chirayath. 2012. "Pre-Training, During-Training and Post-Training Activities as Predictors of Transfer of Training." *The IUP Journal of Management Research*.

Janice Dombrowski. 2024. "Chatbot Examples: How 8 Industries Are Thriving With Chatbots."

Jesse Anglen. 2021. "Top 15 Use Cases Of AI Agents In Business." *RAPID Innovation*.

Jesse Anglen. 2024. "Key Components of Modern AI Agent Architecture." *Rapid innovation*.

Kanbach, Dominik K., Louisa Heiduk, Georg Blueher, Maximilian Schreiter, and Alexander Lahmann. 2024. "The GenAI Is out of the Bottle: Generative Artificial Intelligence from a Business Model Innovation Perspective." *Review of Managerial Science* 18(4): 1189–1220. doi:10.1007/s11846-023-00696-z.

Kaur, Ramanpreet, Dušan Gabrijelčič, and Tomaž Klobučar. 2023. "Artificial Intelligence for Cybersecurity: Literature Review and Future Research Directions." *Information Fusion*. doi:10.1016/j.inffus.2023.101804.

Kavlakoglu, Eda, and Rishi Vaish. 2020. "NLP vs. NLU vs. NLG: The Differences between Three Natural Language Processing Concepts." *IBM*.

Khambholja, Manushi. 2024. "Top 45+ Generative AI Tools for Enterprises and Creatives."

Khurana, Diksha, Aditya Koli, Kiran Khatter, and Sukhdev Singh. 2023. "Natural Language Processing: State of the Art, Current Trends and Challenges." *Multimedia Tools and Applications* 82(3): 3713–44. doi:10.1007/s11042-022-13428-4.

Kumar, Anant. 2023. "Model and Data Versioning: An Introduction to Mlflow and DVC." *Walmart Global Tech Blog*.

LABS, [x]cube. 2024. "Real-Time Generative AI Applications: Challenges and Solutions." *[x]cube LABS*.

LeewayHertz. 2023. "How to Build a Generative AI Solution: A Step-by-Step Guide." *LeewayHertz – AI Development Company*.

Li, Oliver. 2023. "Artificial General Intelligence and Panentheism." *Theology and Science*. doi:10.1080/14746700.2023.2188373.

Lyzr Team. 2024. "Knowledge Graphs." *Lyzr.ai*.

Mah, Pascal Muam, Iwona Skalna, and John Muzam. 2022. "Natural Language Processing and Artificial Intelligence for Enterprise Management in the Era of Industry 4.0." *Applied Sciences (Switzerland)*. doi:10.3390/app12189207.

Marc Dimmick – Churchill Felloww, MMgmt. 2023. "Managing Knowledge in an AI-Driven World."

Mariani, Marcello, and Yogesh K. Dwivedi. 2024. "Generative Artificial Intelligence in Innovation Management: A Preview of Future Research Developments." *Journal of Business Research*. doi:10.1016/j.jbusres.2024.114542.

Marr, Bernard. 2024. "22 Generative AI Workplace Tools And How To Use Them."

Maschler, Benjamin, Hannes Vietz, Hasan Tercan, Christian Bitter, Tobias Meisen, and Michael Weyrich. 2022. "Insights and Example Use Cases on Industrial Transfer Learning." In *Procedia CIRP*, doi:10.1016/j.procir.2022.05.017.

Microsoft. 2022. "Annual Report 2021." *Microsoft*.

Morey, Theodore. 2015. "Customer Data: Designing for Transparency and Trust." *HBR*.

Navdeep Singh Gill. 2024. "The Rise of Multimodal AI Agents: Redefining Intelligent Systems."

Nawaz, Sarfraz. 2024. "AI Agents In Healthcare." *Ampcome*.

Nguyen, Trinh. 2024a. "Traditional AI vs Generative AI: Breaking Down the Basics." *neurond ai*.

Nguyen, Trinh. 2024b. "What Is Generative AI? An In-Depth Look at Machine Creativity." *neurond ai*.

Novac, Pierre Emmanuel, Ghouthi Boukli Hacene, Alain Pegatoquet, Benoît Miramond, and Vincent Gripon. 2021. "Quantization and Deployment of Deep Neural Networks on Microcontrollers." *Sensors*. doi:10.3390/s21092984.

One Beyond. 2024. "Building ChatBots: Powering Business Operations With AI."

Ordax, Eduardo. 2024. "Fine Tuning Vs Pre-Training."

Pan, Dr Jeff, Dr Ernesto Jiménez-Ruiz, Ian Horrocks, and Dr Valentina Tamma. 2024. "Knowledge Graphs."

Pattam, Aruna. 2024. "Agents in Generative AI: A Comprehensive Overview." *MEDIUM*.

Perera, Rivindu, and Parma Nand. 2017. "Recent Advances in Natural Language Generation: A Survey and Classification of the Empirical Literature." *Computing and Informatics*. doi:10.4149/cai_2017_1_1.

Prathima. 2024. "What Is Reinforcement Learning and How Does It Work (Updated 2025)."

Prem, Erich. 2023. "From Ethical AI Frameworks to Tools: A Review of Approaches." *AI and Ethics*. doi:10.1007/s43681-023-00258-9.

PurpleSlate. 2024. "Understanding Artificial General Intelligence (AGI): The Future of AI Technology." *medium*.

RAG, AWS. 2024. "What Is RAG (Retrieval-Augmented Generation)?" *AWS*.

Raiaan, Mohaimenul Azam Khan, Sadman Sakib, Nur Mohammad Fahad, Abdullah Al Mamun, Md. Anisur Rahman, Swakkhar Shatabda, and Md. Saddam Hossain Mukta. 2024. "A Systematic Review of Hyperparameter Optimization Techniques in Convolutional Neural Networks." *Decision Analytics Journal* 11: 100470. doi:10.1016/j.dajour.2024.100470.

Roi Lipman. 2024. "AI Agents: Memory Systems and Graph Database Integration." *FalkprDB*.

Sahitya Arya. 2024. "Top 5 Frameworks for Building AI Agents in 2025." *Analytics vidhya*.

Saleforce. 2024. "What Are AI Customer Service Agents?"

Samant, Rahul Manohar, Mrinal R. Bachute, Shilpa Gite, and Ketan Kotecha. 2022. "Framework for Deep Learning-Based Language Models Using Multi-Task Learning in Natural Language Understanding: A Systematic Literature

Review and Future Directions." *IEEE Access*. doi:10.1109/ACCESS.2022.3149798.

Scott Clark. 2023. "The Evolution of AI Chatbots: Past, Present and Future."

Sengar, Sandeep Singh, Affan Bin Hasan, Sanjay Kumar, and Fiona Carroll. 2024. "Generative Artificial Intelligence: A Systematic Review and Applications." *Multimedia Tools and Applications*. doi:10.1007/s11042-024-20016-1.

Shelf. 2021. "Autonomous AI Agents: The Evolution of Artificial Intelligence." *Shelf*. doi:https://shelf.io/blog/the-evolution-of-ai-introducing-autonomous-ai-agents/.

Sihare, Shyam. 2023. "Artificial Intelligence in Computer Science." In *Advances in Artificial and Human Intelligence in the Modern Era*, doi:10.4018/979-8-3693-1301-5.ch001.

Silva, Marlene, and Daniela Costa. 2023. "Chat GPT and Human Resource Management." *Conferência – Investigação e Intervenção em Recursos Humanos*.

Singh, Loveneet. 2024. "Top 10 AI Agents for Content Creation in 2025: AI Tools & Trends." *Red blink*.

Smith, Blaine E., Amanda Yoshiko Shimizu, Sarah K. Burriss, Melanie Hundley, and Emily Pendergrass. 2025. "Multimodal Composing with Generative AI: Examining Preservice Teachers' Processes and Perspectives." *Computers and Composition* 75: 102896. doi:10.1016/j.compcom.2024.102896.

Statista. 2024. "Generative Artificial Intelligence (AI) Market Size Growth Worldwide from 2021 to 2030." *Statista*.

Steve Denning. 2015. "The 'Jobless Future' Is A Myth." *Forbes*.

Stuart Russell, Peter Norvig. 2020. *Artificial Intelligence: A Modern Approach (Pearson Series in Artifical Intelligence)*.

Systango. 2024. "How Are AI Agents Transforming Global Business Operations?" *systango*.

Takyar, Akash. 2023. "AI for Content Creation: How It Works, Use Cases, Benefits and Solution." *Leeway Hertz*.

Tm, Otto. 2024. "Generative AI: A Productivity Powerhouse with Transformative Potential."

Toward AGI. 2024. "Towards AGI: [Part 1] Agents with Memory." *SuperAGI*.

Tré Wilson. 2023. "Embracing AI in Design: How Designers and AI Can Collaborate." *Fuzzy math*.

Trinh Nguyen. 2024. "Understanding the Concept: What Is an Agent in AI?" *Neurond ei*.

Unclear, Increasingly. 2023. "Storytelling with AI." *medium*.

Wadhwani, Kajol. 2024. "How to Create Your Own Generative AI Solution?" *solulab*.

Wang, Haifeng, Jiwei Li, Hua Wu, Eduard Hovy, and Yu Sun. 2023. "Pre-Trained Language Models and Their Applications." *Engineering*. doi:10.1016/j.eng.2022.04.024.

Wang, Zhe, Jiayi Zhang, Hongyang Du, Ruichen Zhang, Dusit Niyato, Bo Ai, and Khaled B. Letaief. 2024. "Generative AI Agent for Next-Generation MIMO Design: Fundamentals, Challenges, and Vision." : 1–9.

Wilson, Tré. 2023a. "Advice and Insight from the UX Experts at Fuzzy Math." *Fuzzy math*.

Wilson, Tré. 2023b. "Embracing AI in Design: How Designers and AI Can Collaborate." *Fuzzy math*.

Yee, Lareina, Michael Chui, Roger Roberts, and Stephen Xu. 2024. "Why Agents Are the next Frontier of Generative AI." *Quarterly, McKinsey*.

Yuliya Melnik. 2023. "The Top 16 AI Frameworks and Libraries: A Beginner's Guide." *Data camp*.

van der Zant, Tijn, Matthijs Kouw, and Lambert Schomaker. 2013. "Generative Artificial Intelligence." In *Studies in Applied Philosophy, Epistemology and Rational Ethics*, Springer Berlin Heidelberg, 107–20. doi:10.1007/978-3-642-31674-6_8.

Rajiv Kumar, M.S., Director – Data Science

Rajiv Kumar is a seasoned leader in the field of Artificial Intelligence and Data Science, currently serving as the Director of Data Science at Oracle USA. With over 15 years of experience, Rajiv has pioneered innovations in AI platforms, systems design, and scalable solutions that have transformed enterprise-level applications. His expertise spans across a broad range of technologies, including Python, SpaCy, Scikit-Learn, Lang Chain, RAG (Retrieval-Augmented Generation), and cutting-edge AI solutions in machine learning, natural language processing (NLP), and prompt engineering.

Rajiv is recognized for his significant contributions to Oracle's Generative AI Platform-as-a-Service (GenAI PaaS), which has been adopted by thousands of enterprises worldwide. He has also led key projects such as the Skill Taxonomy and Ontology System and AI-powered systems like SR Resolution and the Benefits Advisor platform, aimed at streamlining workflows and enhancing decision-making processes. Under his leadership,

Oracle has earned recognition in the Gartner Magic Quadrant for Human Capital Management (HCM) products.

His work extends beyond the technical, advocating for AI fairness, including bias testing and mitigation. Rajiv's passion for advancing AI technology is reflected in his published research, multiple patents, and industry recognition.

Rajiv's academic background includes a Master's in Information Technology from Denver University and a Bachelor's in Information Technology from Kurukshetra University. Throughout his career, Rajiv has been instrumental in building AI-powered tools for healthcare, workforce analytics, and marketing, driving measurable business outcomes and global impact.

www.ingramcontent.com/pod-product-compliance
Lightning Source LLC
Chambersburg PA
CBHW041302120726
48005CB00014B/1835